Lecture Notes in Bioinformatics 6004

Edited by S. Istrail, P. Pevzner, and M

Editorial Board: A. Apostolico S
T. Lengauer S. Miyano G. Myer.
R. Shamir T. Speed M. Vingron W. w.

Subseries of Lecture Notes in Computer Science

Christian Blaschke   Hagit Shatkay (Eds.)

Linking Literature, Information, and Knowledge for Biology

Workshop of the BioLINK Special Interest Group,
ISMB/ECCB 2009
Stockholm, June 28-29, 2009
Revised Selected Papers

 Springer

Series Editors

Sorin Istrail, Brown University, Providence, RI, USA
Pavel Pevzner, University of California, San Diego, CA, USA
Michael Waterman, University of Southern California, Los Angeles, CA, USA

Volume Editors

Christian Blaschke
Bioalma
C/Ronda de Poniente, 4, 2-C
28760 Tres Cantos, Madrid, Spain
E-mail: blaschke@bioalma.com

Hagit Shatkay
Computational Biology and Machine Learning Lab
School of Computing, Queen's University
Kingston, ON, K7L 3N6 Canada
E-mail: shatkay@cs.queensu.ca

Library of Congress Control Number: 2010927026

CR Subject Classification (1998): J.3, I.2, F.1, I.4, I.5, I.2.7

LNCS Sublibrary: SL 8 – Bioinformatics

ISSN   0302-9743
ISBN-10   3-642-13130-1 Springer Berlin Heidelberg New York
ISBN-13   978-3-642-13130-1 Springer Berlin Heidelberg New York

springer.com

© Springer-Verlag Berlin Heidelberg 2010
Printed in Germany

Typesetting: Camera-ready by author, data conversion by Scientific Publishing Services, Chennai, India
Printed on acid-free paper   06/3180

Preface

This volume of *Lecture Notes in Bioinformatics (LNBI)* contains selected papers from the workshop of the BioLINK Special Interest Group (SIG) on Linking Literature, Information and Knowledge for Biology. The workshop was held June 28–29, 2009 at the ISMB/ECCB 2009 conference in Stockholm (*http://www.iscb.org/ismbeccb2009*).

The BioLINK SIG meeting has been regularly held in association with the ISMB conferences since 2001, focusing on the development and application of resources and tools for biomedical text mining. The SIG is interdisciplinary in nature, and brings together researchers applying natural language processing, text mining, information extraction and retrieval in the biomedical domain, with scientists from bioinformatics and biology. This year the SIG included two special sessions, one on the analysis of images and figures, which play a critial role in scientific papers, and the other on the future of scientific publication. Overview papers concerned with both these topics are included in this volume.

The meeting featured invited talks, presentations of peer-reviewed contributed papers, reports from recent evaluations and workshops, as well as a poster session. The SIG solicited papers that discuss current biological image- and text-related needs, the challenges of meeting them, and the tools that may help address them. Specific topics include: biological image analysis and its relationship to biomedical text mining, application of text-mining tools to biomedical research, automated indexing of the biomedical literature, generation of structured digital abstracts, new evaluation measures to measure utility and usability of text mining tools integrated into the biologist's workflow, as well as new tools and applications, and future directions in biomedical text mining. The publications in this volume were selected from the workshop papers and represent extended versions that have undergone an additional review process by the members of the Program Committee.

We thank the Program Committee for thoroughly reviewing the submitted manuscripts and for their valuable comments that significantly improved the final publications.

December 2009

Christian Blaschke
Hagit Shatkay

Organization

Workshop Program Chairs

Christian Blaschke — Bioalma, Spain

Lynette Hirschman — MITRE Corporation, USA

Scott Markel — Accelrys, USA

Robert Murphy — Carnegie Mellon University, USA

Hagit Shatkay — School of Computing, Queen's University, Canada

Alfonso Valencia — Spanish National Cancer Research Centre, Spain

Program Committee

Alan Aronson — NLM, USA

Olivier Bodenreider — NLM, USA

Monica Chagoyen — Spanish National Centre for Biotechnology, Spain

Aaron Cohen — Oregon Health Sciences University, USA

Kevin B. Cohen — University of Colorado, Health Sciences Center, USA

Mark Craven — University of Wisconsin, Madison, USA

Bob Futrelle — Northeastern University, USA

William Hayes — Biogen IDEC, USA

Martin Krallinger — Spanish National Cancer Research Centre, Spain

Alberto Pascual Montano — Complutense University of Madrid, Spain

Andreas Rechsteiner — University of California, Santa Cruz, USA

Luis Rocha — Indiana University, Bloomington, USA

Karin Verspoor — Los Alamos National Laboratories, USA

John Wilbur — NCBI, NLM, USA

Hong Yu — University of Wisconsin, Milwaukee, USA

Ralf Zimmer — Ludwig-Maximilians University, Munich, Germany

Sponsoring Institutions

We gratefully acknowledge:

The International Society for Computational Biology (ISCB)
The Spanish National Cancer Research Centre
BioCreative II.5 challenge

Table of Contents

BioLINK 2009

Overview of the Ninth Annual Meeting of the BioLINK SIG at ISMB: Linking Literature, Information and Knowledge for Biology

Christian Blaschke[1], Lynette Hirschman[2], Hagit Shatkay[3],
and Alfonso Valencia[4]

[1] Bioalma, Spain
blaschke@almabioinfo.com
[2] MITRE Corporation
lynette@mitre.org
[3] Computational Biology and Machine Learning Lab,
School of Computing, Queen's University, Kingston, Ontario, Canada
shatkay@cs.queensu.ca
[4] Structural Biology and Biocomputing Programme,
Spanish National Cancer Research Centre (CNIO)
valencia@cnio.es

1 About BioLINK

With the increasing availability of textual information related to biological research, such information has become an important component of many bioinformatics applications. Much recent work aims to develop practical tools to facilitate the use of the literature for annotating the vast amounts of molecular data, including gene sequences, transcription profiles and biological pathways. The broad area of biomedical text mining is concerned with using methods from natural language processing, information extraction, information retrieval and summarization to automate knowledge discovery from biomedical text. In the biomedical domain, research has focused on several complex text-based applications, including the identification of relevant literature (information retrieval) for specific information needs, the extraction of experimental findings for assistance in building biological knowledge bases, and summarization – aiming to present key biological facts in a succinct form.

Automated natural language processing (NLP) began in 1947 with the introduction of the idea of machine translation by Warren Weaver, and work on automated (still mechanical) dictionary lookup for translation by Andrew Booth [2,7]. This work was continued throughout the 1950s in research on automatic translation by Bar Hillel, Garvin and others. In the 1950s, work on transformational grammars by Zellig Harris [3], formed the basis for computational linguistics, which was continued by Noam Chomsky, relating natural languages to formal grammars. The field made rapid progress starting in the late 1980s, thanks to a series of conferences focused on evaluation of text mining and information extraction systems: the Message Understanding Conferences (MUCs).

C. Blaschke and H. Shatkay (Eds.): ISBM/ECCB 2009, LNBI 6004, pp. 1–7, 2010.
© Springer-Verlag Berlin Heidelberg 2010

There is also a long history of research on applications of text mining and natural language processing in medicine going back to the late 1960's with Sager's early work on parsing of scientific 'sublanguages' [11,12]. Within biology, text-based methods were introduced in the late 90's. The rapid accumulation of data emerging from advances in sequencing and other high-throughput methods has made the literature a critical source of information for new biological findings. There has been an increasing need for tools to help researchers manage and digest this growing volume of information. Early work in text mining for biology included a 1997 MSc thesis by Timothy Leek [9], and the first publication of an article at the ISMB conference [1].

As is often the case in interdisciplinary fields, communication between the developers of the tools (here – text mining and information extraction tools) and the actual users (in this case – the biologists) is necessary for the development of truly beneficial tools. The BioLINK group was created to address the needs of communication within the field of text mining and information extraction as it is applied to biology and biomedicine, aiming to bring together developers and users. Regular open meetings have been held in association with the ISMB conferences since 2001, facilitating interactions among researchers to exchange ideas with the wider community interested in the latest developments. These meetings focus on the development and the application of resources and tools for biomedical text mining.

BioLINK is interdisciplinary in nature and involves researchers from multiple communities: the users of text mining tools, including curators of biological databases, bench scientists and bioinformaticians; and the researchers who develop methods in natural language processing, ontologies, text mining, image analysis, information extraction and retrieval in order to apply them to problems in the biomedical domain.

In the last decade, BioNLP methods have been maturing, as demonstrated by the results from the BioCreative assessments (Critical Assessment for Information Extraction in Biology). The first BioCreative was held in 2004 [6] and the second in 2007 [8]. During this period, the research community grew (44 team participated in BioCreative II) and the results improved; for instance, in BioCreative II, the best systems achieved almost an f-measure of almost 0.9, in identifying mentions of genes and proteins in text.

Moreover, the availability of the full content of scientific publications has increased to the point where new challenges can be posed and addressed. Recent text mining experiments have begun to use full text articles, in particular to extract information about experimental evidence. Moreover, much of the published material includes images which are of utmost importance for both scientists and database curators. Images typically provide critical supporting evidence for assertions occurring within articles. Image analysis is an important tool in understanding biomedical processes in multiple granularity levels, and also has great potential to enhance document retrieval and categorization. While there has been ongoing interest in developing systems for automatically extracting biological information from the literature, relatively little has been done so far to utilize information from images or on combining text and image data. Recently the challenge of automatically and effectively processing images and figures from the scientific literature is generating much interest in image analysis as a source of biomedical data.[4,10,13,15]

To take into account these new priorities, the scope of BioLINK was extended this year to include the analysis of images and figures within scientific publications. Furthermore, a session about the present and future of scientific publishing was included, in which representatives from scientific journals and community members discussed the impact of information extraction methods (whether from text or images) on producers and consumers of scientific information. Two overview papers, one pertaining to image analysis and the other to the future of scientific publishing, are included in this volume.

Since its inception, it has been part of the mission of the BioLINK SIG to formulate common goals and define standard data sets and uniform evaluation criteria for biomedical text mining systems. In one of the early meetings (in 2002) we proposed to organize an assessment inspired by the well known CASP evaluations for protein structure predictions. This initiative led to the BioCreative challenges that started at the end of 2003. BioCreative is now a well accepted forum for system assessment in the field and has helped to define shared tasks and standardized evaluation criteria, stimulating interaction and exchange of ideas between developers and users of text mining technologies within the biological domain.

2   History of BioLINK Meetings

2001, ISMB Copenhagen, Denmark: Lynette Hirschman and Alfonso Valencia organize the first workshop related to text mining and literature analysis at the ISMB in Copenhagen. This workshop became the predecessor of the BioLINK Special Interest Group that met since then at the annual ISMB conferences.

2002, ISMB Edmonton, Canada: This was the first year where people were invited to contribute publications and present them at the workshop. A wide range of themes were covered including term and entity recognition, augmenting the gene ontology with text mining, functional analysis of genes based on text and literature based discovery.

In addition, interesting discussions took place about the results of the KDD challenge cup [14] and the TREC genomics track [5]. Furthermore, a proposal was presented to organize a new evaluation of text mining systems closer to the needs of biologists. This discussion led to the creation of the BioCreative (Critical Assessment of Information Extraction in Biology) evaluations.

2003, ISMB Brisbane, Australia: The meeting in 2003 focused especially on developing shared infrastructure (tools, corpora, ontologies). The contributed publications discussed corpus resources, extracting protein-protein interactions from text, protein named entity recognition and automatically linking MEDLINE abstracts to the Gene Ontology. At that meeting the first BioCreative was discussed and the initial training data were released.

2004, ISMB Glasgow, Scotland: That year the BioLINK SIG meeting focused on resources and tools for text mining, with special emphasis on the evaluation of these tools. Contributions were in the area of named entity recognition in biomedical texts and infrastructures for term management. The discussions focused on the TREC genomics track and the results of BioCreative.

2005, ISMB Detroit, Michigan: In 2005 the BioLINK meeting was held jointly with the ACL Workshop on Linking Biological Literature, Ontologies and Databases: Mining Biological Semantics. The contributions covered extraction of protein-protein interactions from text, named entity recognition and shallow parsing in biomedical texts, corpora for text mining, functional analysis of genes based on text, and user-oriented biomedical text mining.

2006, ISMB Fortaleza, Brasil: It was decided to organize a joint meeting of the Bio-ontologies and the BioLINK workshops covering both themes. This was done to build on the close relationships between bio-ontologies and biomedical text mining. For example, ontologists apply text-mining techniques to test, check and build ontologies, while the knowledge in ontologies is being used to augment and improve text-mining techniques. The meeting consisted of sessions that focused on the intersection of bio-ontologies and text mining, as well as individual sessions on the use of ontologies in the life sciences and on biomedical text mining.

Authors contributed presentations covering the analysis of cellular processes using text mining, the Protein Description Corpus, corpus annotation guidelines, formats and standards to enhance interoperability of TM systems, (deep) parsing of biomedical text, andfunctional analysis of genes based on text, and the use of images within text.

2007, ISMB Vienna, Austria: Following BioCreative II, the meeting focused on assessments, on standards for annotation both in biological databases and in biomedical text corpora, and on new tools for biomedical text mining.

Papers presented at the meeting discussed both the common themes of named entity recognition and parsing within biomedical texts, along with annotation tools, corpora, extraction of biomedical relationships, and more focus on user-centered text mining systems.

2008, ISMB Toronto, Canada: This BioLINK meeting focused on the theme of automated linkage of the literature to biological resources in support of applications such as: automated indexing of the biomedical literature, generation of structured digital abstracts and the use of text-mined data in biology and bioinformatics pipelines. To stimulate discussion, a forum of "end users" was invited to present their applications and text mining needs, with a specific goal of encouraging partnerships among the end users and the developers of text mining tools.

Papers presented at the workshop covered some of the (by now) traditional topics, such as named entity recognition and parsing of biomedical text, along with more recent topics including corpora and annotation tools, mining information from full text , linking text-based information to biological databases entries.

2009, ISMB Stockholm, Sweden: The latest BioLINK meeting moved beyond text analysis to take into account the analysis of images and figures in scientific publications, in recognition of the key information they provide. Furthermore, the meeting included a session about the future of scientific publishing, and the impact of information extraction methods on producers and consumers of scientific information. This is also the first year in which selected papers from the workshop are being published as conference proceedings within Lecture Notes in Bioinformatics.

Papers presented in this year's meeting discussed named protein-protein interactions, analysis of experimental data and hypothesis generation, linking text to databases entries, text mining systems in support of specific users and applications, augmentation of the gene ontology using text mining, corpus annotation tools, and image analysis in scientific publications.

The following papers, listed here by topic, have been included in this volume:

Effective training of document classifiers in support of database curation

Learning from Positives and Unlabeled Document Retrieval for Curating Bacterial Protein-Protein Interactions by Hongfang Liu, Guixian Xu, Manabu Torii, Zhangzhi Hu, and Johannes Goll.

The authors present a method for training classifiers to detect documents that contain information about protein-protein interactions based on publicly available data. Manually curating training data containing positive and negative examples is time consuming and in many situations not feasible. Often positive examples can be deduced from existing databases, but negative examples are not explicitly given. The authors explore different ways to create reliable negative training data and show that good classifiers can be trained from such automatically created training data.

Toward Computer-Assisted Text Curation: Classification is Easy (Choosing Training Data can be Hard...) by Robert Denroche, Ramana Madupu, Shibu Yooseph, Granger Sutton and Hagit Shatkay.

In this work the authors developed a system to identify abstracts that are likely to describe experimental characterization of proteins, as opposed to abstracts unlikely to contain such information, supporting the curation of characterized proteins. To train and test the classifiers, small hand-curated datasets, as well as a large set based on previously curated abstracts from Swiss-Prot and GO were constructed. The authors show that classifiers trained on relatively small hand-curated datasets perform at a high level very close to the level required by database curators. Another interesting finding of this study was that in more than 80% of the manually examined articles, the abstract and title alone contained sufficient information for determining the relevance of the article for database curation.

Text-mining in support for experimental data analysis

Combining Semantic Relations and DNA Microarray Data for Novel Hypotheses Generation by Dimitar Hristovski, Andrej Kastrin, Borut Peterlin, and Thomas C. Rindflesch.

One important application of text mining systems is to support scientists in interpreting experimental results. Hristovski et al. present a methodology that integrates the results of microarray experiments with a large database of semantic predications extracted by a text mining system from the scientific literature. Examples from microarray data on Parkinson disease are presented to illustrate the way semantic relations shed light on the relationship between current knowledge and information gleaned from the experiment, and help generate novel hypotheses.

Tools

Mining Protein-Protein Interactions from GeneRIFs with OpenDMAP by Andrew D. Fox, William A. Baumgartner Jr., Helen L. Johnson, Lawrence E. Hunter, and Donna K. Slonim.

Standard basic components that can be used as building blocks in biomedical text mining systems are only starting to emerge; during the last few years there are an increasing number of systems have been made available for public use. This work is an example of how publicly available tools for tokenizing, protein named entity recognition and information extraction can be put together to build a system to extract protein interactions from text. The work specifically makes use of a fairly limited and well-structured text, namely, GeneRIFs, and uses the UIMA framework to integrate the different components, thus increasing the possibilities for reuse. Fox et al. describe how the performance of individual processing steps influences the overall results; modules for detecting protein complexes and enhancements to their information extraction patterns are discussed.

Extracting and Normalizing Gene/Protein Mentions with the Flexible and Trainable Moara's Java Library by Mariana L. Neves, José María Carazo and Alberto Pascual-Montano.

The Moara system is an addition to the growing ecosystem of text mining components that are made available to the public and that can help developers focus on specific problems without having to re-implement existing methods. Moara is a Java library that can be easily integrated with other systems or used as a standalone application. It uses machine learning algorithms that are trained to detect and to normalize gene and protein names in the literature.

Integrating images and text

Structured Literature Image Finder: Extracting Information from Text and Images in Biomedical Literature by Luis Pedro Coelho, Amr Ahmed, Andrew Arnold, Joshua Kangas, Abdul-Saboor Sheikh, Eric P. Xing, William W. Cohen, and Robert F. Murphy.

The Structured Literature Image Finder (SLIF) is an example of an advanced, publically available system that can be used by researchers who are not familiar with the underlying technology. It uses a combination of text-mining and image processing to extract information from figures in the biomedical literature. To access the information a web-accessible searchable database is provided to the users. One can query the database for text appearing in figure captions or the images themselves and browse through the publications and their images.

References

1. Andrade, M.A., Valencia, A.: Automatic annotation for biological sequences by extraction of keywords from MEDLINE abstracts. Development of a prototype system. In: Proceedings of the 5th Annual International Conference on Intelligent Systems for Molecular Biology, ISMB 1997 (1997)

2. Chan, S.W.: A Dictionary of Translation Technology. Chinese University Press (2004)
3. Harris, Z.S.: Transfer Grammar. International Journal of American Linguistics 20(4) (October 1954)
4. Hearst, M.A., Divoli, A., et al.: BioText Search Engine: beyond abstract search. Bioinformatics (June 2007)
5. Hersh, W., Bhupatiraju, R.: TREC genomics track overview. In: Proc. of the Twelfth Text Retrieval Conference, TREC 2003 (2003)
6. Hirschman, L., Yeh, A., Blaschke, C., Valencia, A.: Overview of BioCreAtIvE: critical assessment of information extraction for biology. BMC Bioinformatics 6(Suppl. 1), S1 (2005)
7. Hutchins, J.: Warren Weaver Memorandum: 50th Anniversary of Machine Translation. In: MT News International, July 22, pp. 5–6 (1999)
8. Leek, T.R.: Information Extraction Using Hidden Markov Models. Master's thesis, Department of Computer Science, University of California, San Diego (1997)
9. Krallinger, M., et al.: Evaluation of text-mining systems for biology: overview of the Second BioCreative community challenge. Genome Biology 9(Suppl. 2), S1 (2008)
10. Murphy, R.F., Kou, Z., Hua, J., Joffe, M., Cohen, W.W.: Extracting and structuring subcellular location information from on-line journal articles: the Subcellular Location Image Finder. In: Proceedings of IASTED International Conference on Knowledge Sharing and Collaborative Engineering, KSCE 2004 (2004)
11. Sager, N.: Information Reduction of Texts by Syntactic Analysis. Seminar on Computational Linguistics. In: Pratt, A.W., Roberts, A.H., Lewis, K. (eds.) Division of Computer Science and Technology, National Institutes of Health, Bethesda, MD, pp. 46–56 (1966) (PHS Publication No. 1716)
12. Sager, N.: Syntactic Analysis of Natural Language. In: Advances in Computers, vol. 8, pp. 153–188. Academic Press, NY (1967)
13. Shatkay, H., Chen, N., Blostein, D.: Integrating Image Data into Biomedical Text Categorization. Bioinformatics 22(11) (2006); Special issue: Proc. of the Int. Conf. on Intelligent Systems for Molecular Biology (ISMB 2006) (August 2006)
14. Yeh, A.S., Hirschman, L., Morgan, A.A.: Evaluation of text data mining for database curation: lessons learned from the KDD Challenge Cup. In: Proceedings of the 11th Annual International Conference on Intelligent Systems for Molecular Biology, ISMB 2003 (2003)
15. Yu, H., Lee, M.: Accessing Bioscience Images from Abstract Sentences. Bioinformatics 22(11) (2006); Special issue: Proceedings of the 14th Annual International Conference on Intelligent Systems for Molecular Biology (ISMB 2006) (August 2006)

Principles of Bioimage Informatics: Focus on Machine Learning of Cell Patterns

Luis Pedro Coelho[1,2,3], Estelle Glory-Afshar[3,6], Joshua Kangas[1,2,3],
Shannon Quinn[2,3,4], Aabid Shariff[1,2,3], and Robert F. Murphy[1,2,3,4,5,6]

[1] Joint Carnegie Mellon University–University of Pittsburgh Ph.D.
Program in Computational Biology
[2] Lane Center for Computational Biology, Carnegie Mellon University
[3] Center for Bioimage Informatics, Carnegie Mellon University
[4] Department of Biological Sciences, Carnegie Mellon University
[5] Machine Learning Department, Carnegie Mellon University
[6] Department of Biomedical Engineering, Carnegie Mellon University

Abstract. The field of bioimage informatics concerns the development
and use of methods for computational analysis of biological images. Traditionally, analysis of such images has been done manually. Manual annotation is, however, slow, expensive, and often highly variable from one
expert to another. Furthermore, with modern automated microscopes,
hundreds to thousands of images can be collected per hour, making manual analysis infeasible.

This field borrows from the pattern recognition and computer vision literature (which contain many techniques for image processing and
recognition), but has its own unique challenges and tradeoffs.

Fluorescence microscopy images represent perhaps the largest class
of biological images for which automation is needed. For this modality,
typical problems include cell segmentation, classification of phenotypical
response, or decisions regarding differentiated responses (treatment vs.
control setting). This overview focuses on the problem of subcellular location determination as a running example, but the techniques discussed
are often applicable to other problems.

1 Introduction

Bioimage informatics employs computational and statistical techniques to analyze images and related metadata. Bioimage informatics approaches are useful in
a number of applications, such as measuring the effects of drugs on cells [1], local-
izing cellular proteomes [2], tracking of cellular motion and activity [3], mapping
of gene expression in developing embryos [4,5] and adult brains [6], and many

C. Blaschke and H. Shatkay (Eds.): ISBM/ECCB 2009, LNBI 6004, pp. 8–18, 2010.

others. Traditionally, bioimage analysis has been done by visual inspection, but this is tedious and error-prone. Results from visual analysis are not easily compared between papers or groups. Furthermore, as bioimage data is increasingly used to understand gene function on a genome scale, datasets of subtle phenotype changes are becoming too large for manual analysis [7,8,9]. For example, it is estimated that having a single image for every combination of cell type, protein, and timescale would require on the order of 100 billion images [10]. Over the past fourteen years, the traditional visual, knowledge-capture approach has begun to be replaced with automated, data-driven approaches [11,12,13,14,15]. Approaches to quantitatively associate image feature information with additional anatomical and ontological knowledge to generate digital atlases have also been described [16].

This brief review will focus on bioimage informatics approaches to analyzing protein subcellular patterns, with the goal of illustrating many of the principles that are also relevant to other areas of bioimage informatics. The goal of work in this area is to devise a generalizable, verifiable, mechanistic model of cellular organization and behavior that is automatically derived from images [17].

The most commonly used method for determining subcellular location is fluorescence microscopy. Images are collected by tagging a protein or other molecule so that it becomes visible under the fluorescence microscope.

2   Cell Segmentation

A first step in many analysis pipelines is segmentation, which can occur at several levels (e.g., separating nuclei, cells, tissues). This task has been an active field of research in image processing over the last 30 years, and various methods have been proposed and analysed depending on the modality, quality, and resolution of the microscopy images to analyze [18]. We only discuss two commonly used approaches, Voronoi and seeded watershed.

The two first methods require *seed regions* to be defined. These can be simply locally bright regions or can be defined by a more complex procedure. For example, nuclear segmentation, which is a difficult problem by itself [19], is often used to provide seeds for cell-level segmentation.

In Voronoi segmentation [20], a pixel is assigned to the closest seed. This is a very fast method and works well for sparse regions, but does not take into account the location of the cells and makes serious mistakes if the field is crowded.

Seeded watershed segmentation is a region growing approach [21] in which the image can be considered as a landscape with the pixel intensities as elevation. From the seeds, the basins of the landscape are flooded. When two basins are about to merge, a dam is built that represents the boundary between the two cells. This method works well if the seeds are carefully defined (see Figure 1).

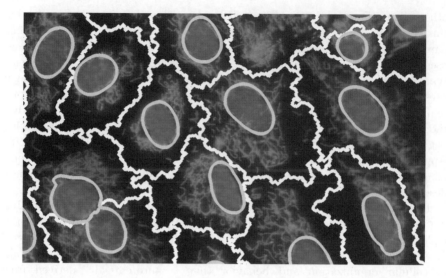

Fig. 1. Seeded Watershed. The nuclear borders (shown in grey) were first identified by a model-based algorithm [22] followed by watershed on the protein channel (resulting in the borders shown in white). Images have been contrast stretched for publication.

3   Supervised Classification

Many applications can be posed as a pattern recognition problem, i.e., given a set of examples of different classes, attempt to group other datapoints into these classes.

Given that even a small image has thousands of pixels, a direct pixel comparison is impossible. Furthermore, two images that differ only in a rigid body motion can have no common pixel values, but represent exactly the same cell state. Thus, the standard approach is to describe the image by a much smaller set of features, where a feature is a numeric function of the image. Once images have been summarized by this smaller set of features, machine learning methods can be applied to learn a classifier.

Features can be computed at several levels: directly from a field that may contain multiple cells, from cell regions (once the image has been segmented as described in the previous section), or from individual subcellular objects in the image. Certain feature classes, such as texture features, are applicable in other vision problems, but features designed specifically for this problem have also been presented (such as those that relate location of the protein of interest to the position of the nucleus) [13].

In some cases, a classifier outperforms a human expert in classification of protein location patterns. The results of one such experiment are shown in Figure 2, where the computer achieves 92% accuracy classifying ten location classes [23], while the human interpreter can only achieve 83% [24]. In another classification study, on the problem of cell detection, it was observed that computers perform

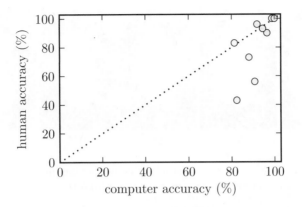

Fig. 2. A comparison of the accuracy in classification indicates that a supervised classification algorithm can perform as well or better than a human expert in recognizing subcellular patterns of ten proteins. Each circle represents one protein.

comparably to a medium-quality expert, but were still outperformed by an experienced expert. However, high variation between experts was observed [25].

4   Shape Analysis

Shapes of cells, nuclei, and organelles are critical for their function. For example, the shape of cells contribute to the overall tissue architecture that have characteristic tissue-specific functions [26]. In pathology, shape is also a useful indicator of deviations from the wild type phenotype. A classical example of a condition that can be identified by the shape of cells is sickle cell anemia, which results in deformed red blood cells. Automated analysis of nuclear shape has received the most attention given the importance of the nucleus in diagnosis.

One approach is to compute a small number of numerical features from the shape image that measure basic properties such as size or the convexity of the shape[1] [27,28].

This feature-based approach is simply an instance of the supervised classification framework as described in Section 3, but approaches that are specifically designed for shape have also been proposed. In particular, diffeomorphic methods have been applied in this area because the shape space of nuclei is non-linear [29]. By capturing how much deformation is needed to morph one shape into another, a distance function in the space of functions is defined. It is then possible to interpolate in the space of shapes [30] to generate intermediate shapes or place a collection of shapes in a low dimensional space in a way that preserves their distance relationships as well as possible. Figure 3 shows deformation distance as a starting nucleus shape is deformed to a target shape.

---

[1] In many applications, it is important to *remove* the influence of size by normalisation, so it depends on the application whether size (or, analogously, orientation) should be captured or normalized out.

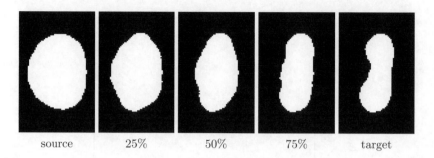

source   25%   50%   75%   target

Fig. 3. The figure shows the transformation using a mapping function from one shape to another. The deformation distance of each image from the starting shape is shown as a percentage of total distance.

Analysis on cell and organelle shapes can be done in a manner similar to nuclear shape. A vast majority of methods use parametric representation of cell shapes [31,32,33].

5   Subcellular Pattern Unmixing

We have described methods to classify fluorescence images according to the depicted subcellular pattern using supervised classification approaches. These methods perform well for defined patterns, but cannot handle patterns that are composed of a mixture. For example, a protein which is partially in the plasma membrane and partially in the nucleus will exhibit a mixed pattern. A system whose model is of discrete assignments of patterns to classes will not be able to even represent the relationship between the *plasma membrane* and *nuclear patterns* and the mixed intermediates. In some situations, defining a mixed class as an extra class might be an acceptable, if inelegant, solution. However, not only does the number of classes grow combinatorially, but if a researcher is interested in quantifying the fraction of fluorescence in each compartment (for example, to study translocation as a function of time), then this solution will not be applicable.

Pattern unmixing directly models this situation. In the *supervised* case, the problem is as follows: the system is given examples of pure patterns and attempts to unmix mixed inputs (i.e., assign mixture fractions for each input condition). In the *unsupervised* form, the system is simply given a collection of images and must identify the fundamental patterns of which mixtures are made.

As one possible approach to this problem, we have described an object-based method for unmixing of subcellular patterns [34]. These methods work on the basis of identifying discrete objects in the image. An object is a contiguous set of pixels that differs in some way from its surroundings (in the simplest case, defined as being above a global threshold value). Patterns are now described by the properties of objects that they show. Mixed patterns will show objects that are characteristic of the basic patterns that compose them. Several variations of these methods have been proposed and this is still an active area of research.

To evaluate the quality of the proposed methods, the correlation between the input mixture coefficients and the underlying fraction of fluorescence in each compartment needs to be computed. In order to know the underlying fractions, Peng et al. [35] built a test set where two dyes that locate differently, but fluoresce similarly were imaged at different concentrations. For supervised unmixing, an 83% correlation with the underlying concentrations was obtained. For the unsupervised case, preliminary results of 72% correlation have been obtained [36].

6   Publicly Available Databases and Analysis of Large Datasets

In previous sections, we described the computational ways in which data is commonly processed. In this section, we give an overview of the sources of high-throughput image data that have been made available by researchers. Some of these have been the subject of automatic analysis, while others have not yet been processed in this way.

Besides the collection of images, these databases also provide annotations, typically manually assigned, describing the protein distribution within cells (see Table 1). The experimental techniques can be grouped into two major families of approaches: fusing the protein of interest to a fluorescence tag, or using fluorescent antibodies that bind to the protein of interest.

Table 1. Examples of microscopy image databases dedicated to subcellular protein location within cells and tissues

Database	Organism	Summary
PSLID-HeLa	H. Sapiens	10 protein distributions tagged by immunofluorescence, 100x magnification
LOCATE-HeLa	H. Sapiens	10 protein distributions tagged by immunofluorescence, 60x magnification
PSLID-RandTag	M. Musculus	ca. 2000 proteins with CD eGFP tagging, 40x magnification
Lifedb	C. Aethiops	ca. 1000 proteins tagged with N- and C-terminal EYFP and ECFP using 1500 cDNA whose vector is expressed in Vero cells, 63x magnification
YPL.db	S. Cerevisiae	371 GFP fusion proteins or vital staining (in 2005)
Yeastgfp	S. Cerevisiae	4,156 strains with detectable GFP signal among the 6,029 GFP positive strains, proteins tagged with GFP *fusion* at their N- and C-terminals, imaged at 100x magnification
HPA-IHC	H. Sapiens	5,000 proteins in 48 normal tissues, 20 cancer tissues, and 47 different cell lines labeled immunohistochemically (DAB and hematoxylin), imaged at 20x
HPA-IF	H. Sapiens	ca. 5,000 proteins in 3 different cell lines (A-431, U-2 OS, and U-251MG) tagged with immunofluorescence, imaged at 63x

In 2002, the first database dedicated to location proteomics, the Protein Subcellular location database (PSLID), was released. Initially composed of 2D single cell images, it has since incorporated various 2D and multicell image datasets. The most recently added dataset represents a collection of images generated by the CD-tagging protocol applied on NIH 3T3 cells [37]. PSLID and its associated software have a unique interface[2] which allows feature calculation, statistical comparison, clustering, classification and the creation of generative models [13].

Given its importance as a model organism, several large collection of yeast images are available. The Yeast Protein Localization database was released in 2002 and 2005 [38,39] (YPL.db[2])[3]. The lack of information on image properties, such as the pixel resolution, limit the potential of automatic analysis of protein distributions in YPL.db[2]. Concurrently, a collection of lines producing GFPfusion proteins for 75% of the open reading frames (ORFs) in S. Cerevisiae was released in 2003[4] [40]. The assignment of each protein into one or a combination of 22 different locations was visually performed. Some fluorescent colocalization experiments with fluorescent markers were made to disambiguate uncertain cases. The same dataset was analyzed by a supervised machine learning approach that showed 81% to 95% accuracy in predicting the correct protein distribution [41]. The accuracy is calculated as the agreement between the labels found by the computer and the hand labeling. However, a closer observation of the mismatches revealed that errors were found in both approaches, human and computer.

The Database for Localization, Interaction, Functional assays and Expression of Proteins (LIFEdb) was created to collect microscopy images of fluorescent fusion proteins produced from human full-length cDNAs expressed in mammalian cell lines [42,43][5]. The dataset was generated by creating fluorescent fusion proteins in Vero cells [44].

The Human Protein Atlas (HPA) project studies the human proteome in situ using antibodies. They collect images of proteins in stained tissues and cell lines in brightfield and fluorescence microscopies [45,46]. They aim to cover the entire human proteome (estimated to consist of ca. 20,000 non-redundant proteins) by 2014. The current release of the HPA database (version 5.0) contains images for more than 8000 antibodies targeting 7000 proteins [46]. The largest collection of images is produced by immunohistochemically (IHC) stained tissue microarrays from normal and cancerous biopsies. The database was extended with 47 different human cell lines commonly used in research and 12 primary blood cells. In addition, a collection of confocal microscopy images of 3 different human cell lines were tagged using immunofluorescence (IF) [47]. The protein distributions were assigned by visual inspection to three different compartments—nuclear, cytoplasm, membranous—for the IHC images. The higher resolution of IF images allows a finer distinction of the protein subcellular distribution into the major cellular organelles. Two machine learning based systems showed good results in generating

---

[2] Available at http://pslid.cbi.cmu.edu/
[3] Available at http://ypl.uni-graz.at/
[4] Available at http://yeastgfp.yeastgenome.org/
[5] Available at http://www.lifedb.de/

labels on both image collections. The classification of the IF images gave 90% accuracy to distinguish 9 subcellular compartments [48] and 81% to distinguish 8 subcellar distribution from non-segmented IHC tissue images [49]. Preliminary results on identifying potential cancer biomarkers by automatically comparing protein distributions in normal and cancer images have also been presented [50].

7 Discussion

This review presented an overview of bioimage informatics, focusing on the problems in analysing fluorescence microscope images.

As a result of the continued progress of bioimage analysis, we increasingly observe domains where automated methods either outperform human analysis or are, at least, comparable with it. Thus, areas of disagreement between the two approaches cannot simply be marked as "classification error on the part of the algorithm." Although humans still outperform computational approaches in general vision tasks by a large margin, the same is not necessarily true in the case of non-natural images of objects which humans do not encounter in their everyday lives (such as fluorescence microscopy images).

Bioimages also generate their own particular problems such as pattern unmixing or learning generative models. These are unique problems that still do not have a definite answer. Biomage informatics still has many active research questions, in the development of entirely new methods to capture information in images, adaptation of existing ones from sister fields, or simply in solving the challenges of applying them to very large collections of data in real time with minimal user intervention. The need for such developments is underscored by the increase in publicly available data which is yet to be fully explored.

References

1. Perlman, Z.E., Slack, M.D., Feng, Y., Mitchison, T.J., Wu, L.F., Altschuler, S.J.: Multidimensional Drug Profiling By Automated Microscopy. Science 306(5699), 1194–1198 (2004)
2. Boland, M.V., Murphy, R.F.: A Neural Network Classifier Capable of Recognizing the Patterns of all Major Subcellular Structures in Fluorescence Microscope Images of HeLa Cells. Bioinformatics 17, 1213–1223 (2001)
3. Meijering, E., Smal, I., Danuser, G.: Tracking in molecular bioimaging. IEEE Signal Processing Magazine 23(3), 46–53 (2006)
4. Peng, H., Myers, E.W.: Comparing in situ mRNA expression patterns of drosophila embryos. In: 8th Intl. Conf. on Computational molecular biology, pp. 157–166 (2004)
5. Zhou, J., Peng, H.: Automatic recognition and annotation of gene expression patterns of fly embryos. Bioinformatics 23(5), 589–596 (2007)
6. Lécuyer, E., Tomancak, P.: Mapping the gene expression universe. Current Opinion in Genetics & Development 18(6), 506–512 (2008)
7. Boland, M.V., Murphy, R.F.: After Sequencing: Quantitative Analysis of Protein Localization. IEEE Engineering in Medicine and Biology Magazine 18(5), 115–119 (1999)

8. Chen, X., Murphy, R.F.: Objective Clustering of Proteins Based on Subcellular Location Patterns. Journal Biomedical Biotechnology 2005(2), 87–95 (2005)
9. Roques, E., Murphy, R.: Objective evaluation of differences in protein subcellular distribution. Traffic 3, 61–65 (2002)
10. Murphy, R.F.: Putting proteins on the map. Nature Biotechnology 24, 1223–1224 (2006)
11. Conrad, C., Erfle, H., Warnat, P., Daigle, N., Lörch, T., Ellenberg, J., Pepperkok, R., Eils, R.: Automatic Identification of Subcellular Phenotypes on Human Cell Arrays. Genome Research 14, 1130–1136 (2004)
12. Gasparri, F., Mariani, M., Sola, F., Galvani, A.: Quantification of the Proliferation Index of Human Dermal Fibroblast Cultures with the ArrayScan High-Content Screening Reader. Journal of Biomolecular Screening 9(3), 232–243 (2004)
13. Glory, E., Murphy, R.F.: Automated Subcellular Location Determination and High Throughput Microscopy. Developmental Cell 12(1), 7–16 (2007)
14. Hamilton, N.A., Pantelic, R.S., Hanson, K., Teasdale, R.D.: Fast automated cell phenotype image classification. BMC Bioinformatics 8, 110 (2007)
15. Huang, K., Lin, J., Gajnak, J., Murphy, R.F.: Image Content-based Retrieval and Automated Interpretation of Fluorescence Microscope Images via the Protein Subcellular Location Image Database. In: IEEE Intl. Symp. Biomedical Imaging, pp. 325–328 (2002)
16. Lein, E., Hawrylycz, M., Ao, N.: Genome-wide atlas of gene expression in the adult mouse brain. Nature 445, 168–176 (2006)
17. Murphy, R.F.: Systematic description of subcellular location for integration with proteomics databases and systems biology modeling. In: IEEE Intl. Symp. Biomedical Imaging, pp. 1052–1055 (2007)
18. Nattkemper, T.W.: Automatic segmentation of digital micrographs: A survey. Studies in health technology and informatics 107(2), 847–851 (2004)
19. Coelho, L.P., Shariff, A., Murphy, R.F.: Nuclear segmentation in microsope cell images: A hand-segmented dataset and comparison of algorithms. In: IEEE Intl. Symp. Biomedical Imaging, pp. 518–521 (2009)
20. Jones, T.R., Carpenter, A.E., Golland, P.: Voronoi-based segmentation of cells on image manifolds. In: Liu, Y., Jiang, T.-Z., Zhang, C. (eds.) CVBIA 2005. LNCS, vol. 3765, pp. 535–543. Springer, Heidelberg (2005)
21. Beucher, S.: Watersheds of functions and picture segmentation. In: IEEE Intl Conf. on Acoustics, Speech and Signal Processing, Paris, pp. 1928–1931 (1982)
22. Lin, G., Adiga, U., Olson, K., Guzowski, J.F., Barnes, C.A., Roysam, B.: A hybrid 3D watershed algorithm incorporating gradient cues and object models for automatic segmentation of nuclei in confocal image stacks. Cytometry Part A 56A(1), 23–36 (2003)
23. Huang, K., Murphy, R.F.: Automated Classification of Subcellular Patterns in Multicell images without Segmentation into Single Cells. In: IEEE Intl. Symp. Biomedical Imaging, pp. 1139–1142 (2004)
24. Murphy, R., Velliste, M., Porreca, G.: Robust Numerical Features for Description and Classification of Subcellular Location Patterns in Fluorescence Microscope Images. Journal of VLSI Signal Processing-Systems for Signal, Image, and Video Technology 35, 311–321 (2003)
25. Nattkemper, T.W., Twellmann, T., Schubert, W., Ritter, H.J.: Human vs. machine: Evaluation of fluorescence micrographs. Computers in Biology and Medicine 33(1), 31–43 (2003)
26. Allen, T.D., Potten, C.S.: Significance of cell shape in tissue architecture. Nature 264(5586), 545–547 (1976)

27. Olson, A.C., Larson, N.M., Heckman, C.A.: Classification of cultured mammalian cells by shape analysis and pattern recognition. Proceedings of the National Academy of Sciences (USA) 77(3), 1516–1520 (1980)
28. Pincus, Z., Theriot, J.A.: Comparison of quantitative methods for cell-shape analysis. Journal of microscopy 227, 140–156 (2007)
29. Rohde, G.K., Ribeiro, A.J.S., Dahl, K.N., Murphy, R.F.: Deformation-based nuclear morphometry: capturing nuclear shape variation in hela cells. Cytometry Part A 73A(4), 341–350 (2008)
30. Peng, T., Wang, W., Rohde, G.K., Murphy, R.F.: Instance-based generative biological shape modeling. In: IEEE Intl. Symp. Biomedical Imaging, vol. 1, pp. 690–693 (2009)
31. Cootes, T.F., Taylor, C.J., Cooper, D.H., Graham, J.: Active shape models—their training and application. Computer Vision and Image Understanding 61(1), 38–59 (1995)
32. Albertini, M.C., Teodori, L., Piatti, E., Piacentini, M.P., Accorsi, A., Rocchi, M.B.L.: Automated analysis of morphometric parameters for accurate definition of erythrocyte cell shape. Cytometry Part A 52A(1), 12–18 (2003)
33. Lehmussola, A., Ruusuvuori, P., Selinummi, J., Huttunen, H., Yli-Harja, O.: Computational framework for simulating fluorescence microscope images with cell populations. IEEE Trans. Medical Imaging 26(7), 1010–1016 (2007)
34. Zhao, T., Velliste, M., Boland, M., Murphy, R.F.: Object type recognition for automated analysis of protein subcellular location. IEEE Trans. on Image Processing 14(9), 1351–1359 (2005)
35. Peng, T., Bonamy, G.M., Glory, E., Daniel Rines, S.K.C., Murphy, R.F.: Automated unmixing of subcellular patterns: Determining the distribution of probes between different subcellular locations. Proceedings of the National Academy of Sciences, USA (2009) (in press)
36. Coelho, L.P., Murphy, R.F.: Unsupervised unmixing of subcellular location patterns. In: Proceedings of ICML-UAI-COLT 2009 Workshop on Automated Interpretation and Modeling of Cell Images (Cell Image Learning), Montreal, Canada (2009)
37. García Osuna, E., Hua, J., Bateman, N.W., Zhao, T., Berget, P.B., Murphy, R.F.: Large-scale automated analysis of location patterns in randomly tagged 3T3 cells. Annals of Biomedical Engineering 35, 1081–1087 (2007)
38. Habeler, G., Natter, K., Thallinger, G.G., Crawford, M.E., Kohlwein, S.D., Trajanoski, Z.: YPL.db: the Yeast Protein Localization database. Nucleic Acids Research 30(1), 80–83 (2002)
39. Kals, M., Natter, K., Thallinger, G.G., Trajanoski, Z., Kohlwein, S.D.: Ypl.db^2: the yeast protein localization database, version 2.0. Yeast 22(3), 213–218 (2005) ·
40. Huh, W.K., Falvo, J.V., Gerke, L.C., Carroll, A.S., Howson, R.W., Weissman, J.S., O'Shea, E.K.: Global analysis of protein localization in budding yeast. Nature 425(6959), 686–691 (2003)
41. Chen, S.C., Zhao, T., Gordon, G., Murphy, R.: Automated image analysis of protein localization in budding yeast. Bioinformatics 23(13), 66–71 (2007)
42. Bannasch, D., Mehrle, A., Glatting, K.H., Pepperkok, R., Poustka, A., Wiemann, S.: LIFEdb: a database for functional genomics experiments integrating information from external sources, and serving as a sample tracking system. Nucleic Acids Research 32, D505–D508 (2004)
43. del Val, C., Mehrle, A., Falkenhahn, M., Seiler, M., Glatting, K.H., Poustka, A., Suhai, S., Wiemann, S.: High-throughput protein analysis integrating bioinformatics and experimental assays. Nucleic Acids Research 32(2), 742–748 (2004)

44. Simpson, J., Wellenreuther, R., Poustka, A., Pepperkok, R., Wiemann, S.: Systematic subcellular localization of novel proteins identified by large-scale cDNA sequencing. EMBO reports 1(3), 287–292 (2000)
45. Uhlen, M., Bjorling, E., Agaton, C., Szigyarto, C.A.K., Amini, B., Andersen, E., Andersson, A.C., Angelidou, P., Asplund, A., Asplund, C., Berglund, L., Bergstrom, K., Brumer, H., Cerjan, D., Ekstrom, M., Elobeid, A., Eriksson, C., Fagerberg, L., Falk, R., Fall, J., Forsberg, M., Bjorklund, M.G., Gumbel, K., Halimi, A., Hallin, I., Hamsten, C., Hansson, M., Hedhammar, M., Hercules, G., Kampf, C., Larsson, K., Lindskog, M., Lodewyckx, W., Lund, J., Lundeberg, J., Magnusson, K., Malm, E., Nilsson, P., Odling, J., Oksvold, P., Olsson, I., Oster, E., Ottosson, J., Paavilainen, L., Persson, A., Rimini, R., Rockberg, J., Runeson, M., Sivertsson, A., Skollermo, A., Steen, J., Stenvall, M., Sterky, F., Stromberg, S., Sundberg, M., Tegel, H., Tourle, S., Wahlund, E., Walden, A., Wan, J., Wernerus, H., Westberg, J., Wester, K., Wrethagen, U., Xu, L.L., Hober, S., Ponten, F.: A Human Protein Atlas for Normal and Cancer Tissues Based on Antibody Proteomics. Molecular & Cellular Proteomics 4(12), 1920–1932 (2005)
46. Berglund, L., Björling, E., Oksvold, P., Fagerberg, L., Asplund, A., Szigyarto, C.A.K., Persson, A., Ottosson, J., Wernérus, H., Nilsson, P., Lundberg, E., Sivertsson, A., Navani, S., Wester, K., Kampf, C., Hober, S., Pontén, F., Uhlén, M.: A genecentric Human Protein Atlas for expression profiles based on antibodies. Molecular & cellular proteomics 7(10), 2019–2027 (2008)
47. Lundberg, E., Sundberg, M., Gräslund, T., Uhlén, M., Svahn, H.A.: A novel method for reproducible fluorescent labeling of small amounts of antibodies on solid phase. Journal of Immunological Methods 322(1-2), 40–49 (2007)
48. Newberg, J., Li, J., Rao, A., Ponten, F., Uhlen, M., Lundberg, E., Murphy, R.F.: Automated analysis of human protein atlas immunofluorescence images. In: IEEE Intl. Symp. Biomedical Imaging, pp. 1023–1026 (2009)
49. Newberg, J., Hua, J., Murphy, R.F.: Location Proteomics: Systematic Determination of Protein Subcellular Location. In: Systems Biology, vol. 500, pp. 313–332. Humana Press (2009)
50. Glory, E., Newberg, J., Murphy, R.F.: Automated comparison of protein subcellular location patterns between images of normal and cancerous tissues. In: IEEE Intl. Symp. Biomedical Imaging, pp. 304–307 (2008)

Summary of the BioLINK Special Interest Group Session on the Future of Scientific Publishing

Scott Markel

Accelrys (SciTegic R&D), San Diego, California, United States of America
smarkel@accelrys.com

Abstract. The 2009 BioLINK SIG (Special Interest Group) meeting held in Stockholm at the ISMB conference (Intelligent Systems for Molecular Biology—the annual conference of the International Society for Computational Biology) included a new session devoted to the future of scientific publishing. The session was added in response to the very favorable reviews of the 2008 Special Session on the same topic.

1 Introduction

The session's focus on scientific publishing fit in well with BioLINK's coverage of natural language processing, text mining, and information extraction and retrieval in the biomedical domain. The publishing session was co-organized by the BioLINK SIG with the collaboration of the ISCB Publications Committee (http://www.iscb.org/iscb-leadership-a-staff-/117) and *PLoS Computational Biology* (http://www.ploscompbiol.org). The session format was expanded to two two-hour segments, both of which were open to ISMB conference registrants. The first segment featured scientific presentations from David Shotton, Anita de Waard, Dietrich Rebholz-Schuhmann, and Philip E. Bourne. The second segment included presentations from journal publishers and finished with an open discussion.

"Adventures in Semantic Publishing: Exemplar Semantic Enhancements of a Research Article"

David Shotton (University of Oxford)

In 2008 Shotton undertook manual semantic enhancements to a biomedical research article, providing enrichment to its content and increased access to datasets within it, to provide a compelling existence proof of the possibilities of semantic publication (http://dx.doi.org/10.1371/journal.pntd.0000228.x001). These semantic enhancements included provision of live DOIs and hyperlinks; semantic markup of textual terms with links to relevant third-party information resources; interactive figures; a reorderable reference list; a document summary containing a study summary, a tag cloud, and a citation analysis; and two novel types of semantic enrichment: the first a Supporting Claims Tooltip to permit "Citations in Context", and the second Tag Trees that bring together semantically related terms. In addition, he published downloadable

C. Blaschke and H. Shatkay (Eds.): ISBM/ECCB 2009, LNBI 6004, pp. 19–22, 2010.
© Springer-Verlag Berlin Heidelberg 2010

spreadsheets containing data from within tables and figures, enriched these with provenance information, and demonstrated various types of data fusion (mashups) with results from other research articles and with Google Maps. Shotton also published machine-readable RDF metadata both about the article and about the references it cites, for which we developed a Citation Typing Ontology, CiTO (http://purl.org/net/cito/).

In his presentation, he explained what was achieved by means of a live link to the online enhanced paper, discussed the significance of this work in terms of recent developments in automated text mining, and considered the future of semantic publishing as part of mainstream research journal production workflows.

"From Proteins to Hypotheses—Some Experiments in Semantic Enrichment"

Anita de Waard (Elsevier Labs, Amsterdam, and Utrecht Institute of Linguistics, Utrecht University)
deWaard discussed a number of initiatives about improving and enhancing access to scientific knowledge from collections of research articles. First, at Elsevier Labs, she added manually annotated Structured Digital Abstracts in *FEBS Letters* articles (http://www.febsletters.org/content/sda_summary) containing curated data on protein–protein interactions. To help authors identify these, within the OKKAM EU Project she is creating a Word plug-in using text mining technologies connected by a Web Service to the authoring environment. She also discussed work at Utrecht University regarding scientific discourse analysis, focusing on the identification of different cognitive realms (experiments and conceptual models) in a full-text research publication, and the linguistic methods by which authors identify the epistemic ("truth value") status of statements. She then discussed some collaborative efforts for the creation of a common framework to bootstrap efforts in this area. de Waard concluded by describing efforts at Elsevier Labs and the University of Utrecht to stimulate and contribute to the discussion on changing models of publishing. She helped organized the Elsevier Grand Challenge (http://www.elseviergrandchallenge.com/) to help stimulate collaboration with researchers interested in addressing the redefinition of scientific communication.

"ELIXIR Scientific Literature Interdisciplinary Interactions"

Dietrich Rebholz-Schuhmann (European Bioinformatics Institute)
Scientific literature is nowadays distributed in electronic form through online Web portals. ELIXIR Work Package 8 (WP8; http://www.elixir-europe.org/page.php?page=wp8) analyzes the academic and commercial stakeholders' needs for automatic exploitation of the resources.

Scientific literature is kept in national and international repositories that currently still lack connectivity. The biomedical community is driven by the idea of the integration of all data resources (including literature) from the level of molecular biology to medicine, leading to multidisciplinary research. The appropriate infrastructure and tools need to be in place to facilitate full exploitation of the literature across scientific domains and at various levels of end user expertise. Scientific literature is unstructured in contrast to the scientific databases. This has led to (1) the development of text mining and knowledge discovery solutions that recover facts from the scientific literature, (2) curation efforts to include scientific facts into the main databases, and (3) efforts around various wiki-like projects to produce annotations. The exploitation of the scientific literature has to (1) fulfill multidisciplinary needs, (2) exploit ontological resources (Semantic Web approaches), (3) deliver enhanced digital content, and (4) follow standards for efficient integration.

"OpenID vs. ResearcherID"

Philip E. Bourne (University of California San Diego)
Scientists (at least their profiles) and their scholarly output exist in cyberspace, but the relationship between the two is far from established. Scientists may not be identified uniquely, and much of their output is not easily referenced. The Digital Object Identifier (DOI) was a big step in uniquely identifying a scientific journal publication, and has been embraced by the majority of publishers. Bourne proposed that the time had come for extending this scheme to uniquely identify scientists (authors) with all their respective scholarly output. This is much more than traditional journal publications, and includes database depositions, reviews for grants and journals, blog postings: in fact anything they wish to have uniquely associated with their name. He discussed efforts in this direction and what he think it will take to really make such a scheme work—a scheme that starts with the publishers.

2 Panel Discussion

The publishers' panel followed the scientific presentations. The publishers were free to comment on the presentations or to address other topics, such as validation processes and quality measures (e.g., the future of the peer review model, alternatives to impact factors), dissemination (e.g., open-access models), and discoverability (e.g., linking, applying new technologies). Participants included Claire Bird (Oxford University Press), Mark Patterson (Public Library of Science), Matt Day (*Nature*), Robert Campbell (Wiley-Blackwell), Matt Cockerill (BioMed Central), David Tranah (Cambridge University Press), and Anita de Waard (Elsevier).

Bird said OUP journals are adding semantic technologies and Web 2.0 technology. She sees publishers adding value in the areas of quality control and archiving so that scientific content can be searched and extracted. She mentioned OpenWetware and WikiGenes as projects where articles can be annotated.

Patterson said most publishers still focus on citations as the means for tacking usage even though other mechanisms are available. PLoS now provides metrics for each article.

Day focused his comments on experimental data associated with articles, especially large data sets. He mentioned that this would need to fit in with data archives like those provided by European Molecular Biology Laboratory (EMBL) and National Center for Biotechnology Information (NCBI).

Campbell is CrossRef's chairman and addressed versioning of articles. How does a reader know that a given article version is the "final" one? CrossRef will be using CrossMark, an authentication logo, to indicate the current published version.

Cockerill also addressed the problems of experimental data, making it accessible and usable. BMC is starting the Journal of Biomedical Semantics. Dietrich Rebholz-Schuhmann and Goran Nenadic are the editors.

Tranah commented on the impact of open access publishing on traditional publishing and proposed that impact factors are often not very good indicators of a journal's or a paper's influence. He gave examples across several domains showing the wide variability of impact factors for the best journals in those fields.

An open discussion with the panelists and session attendees concluded the session.

Structured Literature Image Finder: Extracting Information from Text and Images in Biomedical Literature

Luís Pedro Coelho[1,2,3], Amr Ahmed[4,5], Andrew Arnold[4], Joshua Kangas[1,2,3], Abdul-Saboor Sheikh[3], Eric P. Xing[1,2,3,4,5,6], William W. Cohen[1,2,3,4], and Robert F. Murphy[1,2,3,4,6,7]

[1] Lane Center for Computational Biology, Carnegie Mellon University
[2] Joint Carnegie Mellon University–University of Pittsburgh Ph.D. Program in Computational Biology
[3] Center for Bioimage Informatics, Carnegie Mellon University
[4] Machine Learning Department, Carnegie Mellon University
[5] Language Technologies Institute, Carnegie Mellon University
[6] Department of Biological Sciences, Carnegie Mellon University
[7] Department of Biomedical Engineering, Carnegie Mellon University

Abstract. SLIF uses a combination of text-mining and image processing to extract information from figures in the biomedical literature. It also uses innovative extensions to traditional latent topic modeling to provide new ways to traverse the literature. SLIF provides a publicly available searchable database (http://slif.cbi.cmu.edu).

SLIF originally focused on fluorescence microscopy images. We have now extended it to classify panels into more image types. We also improved the classification into subcellular classes by building a more representative training set. To get the most out of the human labeling effort, we used active learning to select images to label.

We developed models that take into account the structure of the document (with panels inside figures inside papers) and the multi-modality of the information (free and annotated text, images, information from external databases). This has allowed us to provide new ways to navigate a large collection of documents.

1 Introduction

Thousands of papers are published each day in the biomedical domain. Working scientists therefore struggle to keep up with all the results that are relevant to them. Traditional approaches to this problem have focused solely on the text of papers. However, images are also very important as they often contain the primary experimental results being reported. A random sampling of such figures in the publicly available PubMed Central database reveals that in some, if not most of the cases, a biomedical figure can provide as much information as a normal abstract. Thus, researchers in the biomedical field need automated systems that can help them find information quickly and satisfactorily. These

C. Blaschke and H. Shatkay (Eds.): ISBM/ECCB 2009, LNBI 6004, pp. 23–32, 2010.
© Springer-Verlag Berlin Heidelberg 2010

systems should provide them with a structured way of browsing the otherwise unstructured knowledge in a way that inspires them to ask questions that they never thought of before.

Our team developed the first system for automated information extraction from images in biological journal articles (SLIF, the "Subcellular Location Image Finder," first described in 2001 [1]). Since then, we have reported a number of improvements to the SLIF system [2,3,4]. In part reflecting this, we are rechristening SLIF as the "Structured Literature Image Finder."

Most recently, we have added support for more image types, improved classification methods, and added features based on multi-modal latent topic modeling. Topic modeling allows for innovative user-visible features such as "browse by topic," retrieval of topic-similar images or figures, or interactive relevance feedback. Traditional latent topic approaches have had to be adapted to the setting where documents are composed of free and annotated text and images arranged in a structured fashion. We have also added a powerful tool for organizing figures by topics inferred from both image and text, and have provided a new interface that allows browsing through figures by their inferred topics and jumping to related figures from any currently viewed figure. We have performed a user study where we asked users to perform typical tasks with SLIF and report whether they found the tool to be useful. The great majority of responses were very positive [5].

SLIF provides both a pipeline for extracting structured information from papers and a web-accessible searchable database of the processed information. Users can query the database for various information appearing in captions or images, including specific words, protein names, panel types, patterns in figures, or any combination of the above.

2   Overview

The SLIF processing pipeline is illustrated in Figure 1. After preprocessing, image and caption processing proceed in parallel. The results of these two modules then serve as input to the topic modeling framework.

The first step in image processing is to split the image into its panels, then identify the type of image in each panel. If the panel is a fluorescence micrograph image (FMI), the depicted subcellular localization is automatically identified [1]. In addition, panel labels are identified through optical character recognition, and scale-bars, if present, are identified. Annotations such as white arrows are removed.

In parallel, the caption is parsed and relevant biological entities (protein and cell types) are extracted from the caption using named entity recognition techniques. Also, the caption is broken up into logical scopes (sub-captions, identified by markers such as "(A)"), which will be subsequently linked to panels.

The last step in the pipeline aggregates the results of image and caption processing by using them to infer underlying themes in the collection of papers. These are based on the free text in the caption, on the annotated text

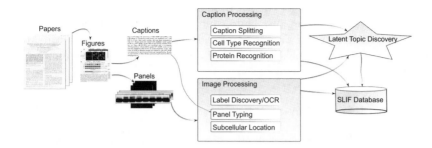

Fig. 1. Overview of SLIF pipeline

(i.e., protein and cell type names are not processed as simple text), and the image features and subcellular localization. This results in a low-dimensional representation of the data, which is used to implement retrieval by example ("find similar papers") or interactive relevance feedback navigation.

Access to the results of this pipeline is provided via a web interface or pro-gramatically with SOAP queries. Results presented always link back to the full paper for user convenience.

3 Caption Processing

A typical caption, taken from [6], reads as:

> S1P induces relocalization of both **p130Cas** and **MT1-MMP** to pe-ripheral **actin**-rich structures. (**A**) **HUVEC** were stimulated for 15 min with 1 μM S1P and stained with polyclonal **MT1-MMP** [...]. (**B**) Cells were stimulated with S1P as described above [...]. Scale bars are **10μm**.

We have highlighted, in bold, the pieces of information which are of interest to SLIF. The text contains both a global portion (the first sentence) and portions scoped to particular panels (marked by "(A)" and "(B)"). Thus the caption is broken up into three parts, one global, and two specific to a panel. In order to un-derstand what the image represents, SLIF extracts the names of proteins present (p130Cas, MT1-MMP,...), as well as the cell line (HUVEC) using techniques described previously. Additionally, SLIF extracts the length(s) of any scale bars to be associated with scale bars extracted from the image itself.

The implementation of this module is described in greater detail else-where [2,4,5,7].

4 Image Processing

4.1 Figure Splitting

The first step in our image processing pipeline is to divide the extracted figures into their constituent components, since in a majority of the cases, the figures are comprised of multiple panels to depict similar conditions, corresponding analysis, etc. For this purpose, we employ a figure-splitting algorithm that recursively finds constant-intensity boundary regions to break up the image hierarchically. Large regions are considered panels, while small regions are most likely annotations. This method that was previously shown to perform well [1].

4.2 "Ghost" Detection

FMI panels are often false color images composed of related channels. However, due to treatment of the image for publication or compression artifacts, it is common that an image that contains one or two logical colors (and is so perceived by the human reader), will have signal in all three color channels. The extra channel, we call a "ghost" of the signal-carrying channels. Figure 2 illustrates this phenomenon.

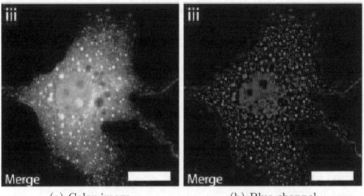

(a) Color image   (b) Blue channel

Fig. 2. Example of a ghost image. Although the color image is obviously a two-channel image (red and green), there is a strong bleed-through into the blue component.

Algorithm 1. Ghost Detection Algorithm

---

1 White := pixelwise-min(R,G,B)
2 M := max(R−White, G−White, B−White)
3 **foreach** $ch \in (R,G,B)$ **do**
4 | Residual := ch−White
5 | sort pixels from Residual
6 | **if** *95% percentile pixel* $< 10\%M$ **then**
7 | |_   ch is a ghost

To detect ghosts, we first compute the white component of the image, i.e., the pixel-wise minimum of the 3 channels. We then subtract this component from each channel so that the regions with homogeneous intensities across all channels (e.g. annotations or pointers) get suppressed. Then, for each channel, we verify if its 95%-percentile pixel is at least 10% of the overall highest pixel value. These two values were found empirically to reject almost all ghosts, with a low rate of false negatives (a signal carrying channel that has less than 5% bright pixels will be falsely rejected, but we found the rate of false positives to be low enough to be acceptable). Algorithm 1 illustrates this process in pseudo-code.

4.3   Panel Type Classification

SLIF was originally designed to process only FMI panels. Recently, we expanded the classification to other panel types, in a way similar to other recent systems [8,9,10].

Panels are classified into one of six panel classes: (1) FMI, (2) gel, (3) graph or illustration, (4) light microscopy, (5) X-ray, or (6) photograph. To build a training set for this classification problem, while minimizing labeling effort, we used empirical risk reduction, an active learning algorithm [11]. We used a libSVM-based classifier as the base algorithm. In order to speed up the process, at each round, we labeled the 10 highest ranked images plus 10 randomly selected images. The process was seeded by initially labeling 50 randomly selected images. This resulted in ca. 700 labeled images.

The previous version of SLIF already had a good FMI classifier, which we have kept. Given its frequency and importance, we focused on the *gel* class as the next important class. Towards this goal, we define a set of features based on whether certain marker words appeared in the caption that would signal gels[1] as well as a set of substrings for the inverse class[2]. A classifier based on these boolean features was learned using the ID3 decision tree algorithm [12] with precision on the positive class as the function being maximized. This technique was shown, through 10 fold cross-validation, to obtain very high precision (91%) at the cost of moderate recall (66%). Therefore, examples considered positive are labeled as such, but examples considered negative are passed on to a classifier based on image features. In addition to the features developed for FMI classification, we measure the fraction of variance that remains in the image formed by the differences between horizontally adjacent pixels:

$$h(I) = \frac{\text{var}(I_{y,x-1} - I_{y,z})}{\text{var}(I_{y,x})}. \tag{1}$$

Gels, consisting of horizontal bars, score much lower on this measure than other types of images. Furthermore, we used 26 Haralick texture features [13]. Images were then classified into the six panel type classes using a support vector machine (SVM) based classifier. On this system, we obtain an overall accuracy of 69%.

---

[1] The positive markers were: *Western, Northern, Southern, blot, lane, RT* (for "reverse transcriptase"), *RNA, PAGE, agarose, electrophoresis*, and *expression*.

[2] The negative markers were: *bar* (for bar charts), *patient, CT*, and *MRI*.

Therefore, the system proceeds through 3 classification levels: the first level classifies the image into FMI or non-FMI using image based features; the second level uses the textual features described above to identify gels with high-precision; finally, if both classifiers gave negative answers, an SVM operating on image-based features does the final classification.

4.4   Subcellular Location Pattern Classification

Perhaps the most important task that SLIF supports is to extract information based on the subcellular localization depicted in FMI panels.

To provide training data for pattern classifiers, we hand-labeled a set of images into four different subcellular location classes: (1) *nuclear*, (2) *cytoplasmic*, (3) *punctate*, and (4) *other*, following the active learning methodology described above for labeling panel types. The active learning loop was seeded using images from a HeLa cell image collection that we have previously used to demonstrate the feasibility of automated subcellular pattern classification [14].

The dataset was filtered to remove images that, once thresholded using the methods we described previously [14], led to less than 80 above-threshold pixels, a value which was empirically determined. This led to the rejection of 4% of images. In classification, if an image meets the rejection criterion, it is assigned into a special *don't know* class.

We computed previously described field-level features to represent the image patterns (field-level features are features that do not require segmentation of images into individual cell regions). We added a new feature for the size of the median object (which is a more robust statistic than the previously used mean object size). Experiments using stepwise discriminant analysis as a feature selection algorithm [15] showed that this was an informative feature. If the scale is inferred from the image, then we normalize this feature value to square microns. Otherwise, we assume a default scale of $1\mu m$/pixel.

We also adapted the threshold adjacency statistic features (TAS) from Hamilton et al. [16] to a parameter-free version. The original features depended on a manually controlled-two-step binarization of the image. For the first step, we use the Ridler–Calvard algorithm to identify a threshold instead of a fixed threshold [17]. The second binarization step involves finding those pixels that fall into a given interval such as $[\mu - M, \mu + M]$, where μ is the average pixel value of the above-threshold pixel and M is a margin (set to 30 in the original paper). We set M to the standard deviation of the above threshold pixels.[3] We call these *parameter-free* TAS.

On the 3 main classes (Nuclear, Cytoplasmic, and Punctate), we obtained 75% accuracy (as before, reported accuracies are estimated using 10 fold cross-validation and the classifier used was an svm). On the four classes, we obtained 61% accuracy.

---

[3] Other methods for binarizing the image presented by Hamilton et al. are handled analogously.

4.5   Panel and Scope Association

As discussed above, figures are composed of a set of panels and a set of subimages which are too small to be panels. To associate panels with their caption pointers (e.g., identifying which panel is panel "A" if such a mention is made in the caption), we parse all panels and other sub-images using optical character recognition (OCR). In the simple case, the panel contains the panel annotation and there is a one-to-one match to annotations in the caption. Otherwise, we match panels to the nearest found in-image annotation.

5   Topic Discovery

The previous modules result in panel-segmented, structurally and multi-modally annotated figures: each figure is composed of multiple panels, and the caption of the whole figure is parsed into scoped caption, global caption, and protein entities. Each scoped caption is associated with a single panel and the global caption is shared across panels and provides contextual information. Given this organization, we would like to build a system for querying across modality and granularity. For instance, the user might want to search for biological figures given a query composed of key words and protein names (across-modality), or the user might want to retrieve figures similar to a given panel (across-granularity) or a given other figure of interest. In this section, we describe our approach to address this problem using topic models.

Topic models aim towards discovering a set of latent themes present in the collection of papers. These themes are called topics and serve as the basis for visualization and semantic representation. Each topic k consists of a triplet of distributions: a multinomial distribution over words β_k, a multinomial distribution of protein entities Ω_k, and a gaussian distribution over every image feature s, $(\mu_{k,s}, \sigma_{k,s})$. Given these topics, a graphical model is defined that generates figure f given these topics (see [18] for a full description). There are two main steps involved in building our topic model: inference and learning. In learning, given a set of figures, the goal is to learn the set of topics $(\beta_k, \Omega_k, \{\mu_{k,s}, \sigma_{k,s}\})$ that generates the collection using Bayesian inference [18]. On the other hand, given the discovered topics and a new figure f, the goal of inference is to deduce the latent representation of this figure $\theta_f = (\theta_{f,1} \cdots \theta_{f,k})$, where the component $\theta_{f,k}$ defines how likely topic k will appear in figure f. Moreover, for each panel p in figure f, the inference step also deduces its latent representation: $\theta_{f,p} = (\theta_{f,p,1} \cdots \theta_{f,p,k})$. In addition, from the learning step, each word w and protein entity r can also be represented as a point in the topic space: $\theta_w = (\beta_{1,w}, \cdots, \beta_{k,w})$ and $\theta_r = (\Omega_{1,r}, \cdots, \Omega_{k,r})$.

This results in a unified space where each figure, panel, word and protein entity is described using a point in this space which facilitates querying across modality and granularity. For instance, given a query $q = (w_1, \cdots, w_n, r_1, \cdots, r_m)$ composed of a set of text words and protein entities, we can rank figures according to this query using the query language model [19] as follows:

$$P(q|f) = \prod_{w \in q} P(w|f) \prod_{r \in q} P(r|f) = \prod_{w \in q} \left[\sum_k \theta_{f,k} \beta_{k,w} \right] \prod_{r \in q} \left[\sum_k \theta_{f,k} \Omega_{k,r} \right]$$
$$= \prod_{w \in q} \left[\theta_f \odot \theta_w \right] \prod_{r \in q} \left[\theta_f \odot \theta_r \right] \tag{2}$$

Equation 2 is a simple dot product operation between the latent representations of each query item and the latent representation of the figure in the induced topical space. The above measure can then be used to rank figures for retrieval. Moreover, given a figure of interest f, other figures in the database can be ranked based on similarity to this figure as follows:

$$sim(f'|f) = \sum_k \theta_{f,k} \theta_{f',k} = \theta_f \odot \theta_{f'} \tag{3}$$

In addition to the above capabilities, the discovered topics endow the user with a bird's eye view over the paper collection and can serve as the basis for visualization and structured browsing. Each topic f summarizes a theme in the collection and can be represented to the user along three dimensions: top words (having high values of $\beta_{k,w}$), top proteins entities (having high values of $\Omega_{k,r}$), and a set of representative panels (panels with high values of $\theta_{f,p,k}$). Users can decide to display all panels (figures) that are relevant to a particular topic of interest [5,18].

6   Discussion

We have presented a new version of SLIF, a system that analyzes images and their associated captions in biomedical papers. SLIF demonstrates how text-mining and image processing can intermingle to extract information from scientific figures. Figures are broken down into their constituent panels, which are handled separately. Panels are classified into different types, with the current focus on FMI and gel images. FMIs are further processed by classifying them into their depicted subcellular location pattern. The results of this pipeline are made available through either a web-interface or programmatically using SOAP technology.

A new addition to our system is latent topic discovery which is performed using both text and image. This is based on extending traditional models to handle the structure of the literature and allows us to customize these models with domain knowledge (by integrating the subcellular localization looked up from a database, we can see relations between papers using knowledge present outside of them).

Our most recent human-labeling efforts (of panel types and subcellular location) were performed using active learning to extract the most out of the human effort. We plan to replicate this approach in the future for any other labeling effort (e.g., adding a new collection of papers). Our current labeling efforts were necessary to collect a dataset that mimicked the characteristics of the task at hand (images from published literature) and improve on our previous use of datasets that did not show all the variations present in real published datasets. These datasets are available for download from the SLIF webpage (http://slif.cbi.cmu.edu) so that they can be used by other system developers and for building improved pattern classifiers.

Acknowledgments

The SLIF project is currently supported by NIH grant R01 GM078622. L.P.C. was partially supported by a grant from the Fundação Para a Ciência e Tecnologia (grant SFRH/BD/37535/2007).

References

1. Murphy, R.F., Velliste, M., Yao, J., Porreca, G.: Searching online journals for fluorescence microscope images depicting protein subcellular location patterns. In: BIBE 2001: Proceedings of the 2nd IEEE International Symposium on Bioinformatics and Bioengineering, Washington, DC, USA, pp. 119–128. IEEE Computer Society, Los Alamitos (2001)
2. Cohen, W.W., Wang, R., Murphy, R.F.: Understanding captions in biomedical publications. In: KDD 2003: Proceedings of the ninth ACM SIGKDD international conference on Knowledge discovery and data mining, pp. 499–504. ACM, New York (2003)
3. Murphy, R.F., Kou, Z., Hua, J., Joffe, M., Cohen, W.W.: Extracting and structuring subcellular location information from on-line journal articles: The subcellular location image finder. In: Proceedings of IASTED International Conference on Knowledge Sharing and Collaborative Engineering, pp. 109–114 (2004)
4. Kou, Z., Cohen, W.W., Murphy, R.F.: A stacked graphical model for associating sub-images with sub-captions. In: Altman, R.B., Dunker, A.K., Hunter, L., Murray, T., Klein, T.E. (eds.) Proceedings of the Pacific Symposium on Biocomputing, pp. 257–268. World Scientific, Singapore (2007)
5. Ahmed, A., Arnold, A., Coelho, L.P., Kangas, J., Sheikh, A.S., Xing, E.P., Cohen, W.W., Murphy, R.F.: Structured literature image finder: Parsing text and figures in biomedical literature. Journal of Web Semantics (2009) (in press)
6. Gingras, D., Michaud, M., Tomasso, G.D., Bliveau, E., Nyalendo, C., Bliveau, R.: Sphingosine-1-phosphate induces the association of membrane-type 1 matrix metalloproteinase with p130cas in endothelial cells. FEBS Letters 582(3), 399–404 (2008)
7. Kou, Z., Cohen, W.W., Murphy, R.F.: High-recall protein entity recognition using a dictionary. Bioinformatics 21, i266–i273 (2005)
8. Geusebroek, J.M., Hoang, M.A., van Gernert, J., Worring, M.: Genre-based search through biomedical images. In: Proceedings of 16th International Conference on Pattern Recognition, vol. 1, pp. 271–274 (2002)
9. Shatkay, H., Chen, N., Blostein, D.: Integrating image data into biomedical text categorization. Bioinformatics 22(14), 446–453 (2006)

10. Rafkind, B., Lee, M., Chang, S., Yu, H.: Exploring text and image features to classify images in bioscience literature. In: Proceedings of the BioNLP Workshop on Linking Natural Language Processing and Biology at HLT-NAACL, Morristown, NJ, USA. Association for Computational Linguistics, pp. 73–80 (2006)
11. Roy, N., Mccallum, A.: Toward optimal active learning through sampling estimation of error reduction. In: Proc. 18th International Conf. on Machine Learning, pp. 441–448. Morgan Kaufmann, San Francisco (2001)
12. Mitchell, T.M.: Machine Learning. McGraw-Hill, New York (1997)
13. Haralick, R.M.: Statistical and structural approaches to texture. Proceedings of the IEEE 67, 786–804 (1979)
14. Boland, M.V., Murphy, R.F.: A neural network classifier capable of recognizing the patterns of all major subcellular structures in fluorescence microscope images of HeLa cells. Bioinformatics 17(12), 1213–1223 (2001)
15. Jennrich, R.: Stepwise Regression & Stepwise Discriminant Analysis. In: Statistical Methods for Digital Computers, pp. 58–95. John Wiley & Sons, Inc., New York (1977)
16. Hamilton, N., Pantelic, R., Hanson, K., Teasdale, R.: Fast automated cell phenotype image classification. BMC Bioinformatics 8(1), 110 (2007)
17. Ridler, T., Calvard, S.: Picture thresholding using an iterative selection method. IEEE Trans. Systems, Man and Cybernetics 8(8), 629–632 (1978)
18. Ahmed, A., Xing, E.P., Cohen, W.W., Murphy, R.F.: Structured correspondence topic models for mining captioned figures in biological literature. In: Proceedings of The Fifteenth ACM SIGKDD International Conference on Knowledge Discovery and Data Mining (KDD 2009), pp. 39–47. ACM, New York (2009)
19. Ponte, J.M., Croft, W.B.: A language modeling approach to information retrieval. In: SIGIR 1998: Proceedings of the 21st annual international ACM SIGIR conference on Research and development in information retrieval, pp. 275–281. ACM, New York (1998)

Toward Computer-Assisted Text Curation:
Classification Is Easy
(Choosing Training Data Can Be Hard...)

Robert Denroche[1], Ramana Madupu[2], Shibu Yooseph[2],
Granger Sutton[2], and Hagit Shatkay[1]

[1] Computational Biology and Machine Learning Lab,
School of Computing, Queen's University, Kingston, Ontario, Canada
{Denroche,Shatkay}@cs.queensu.ca
[2] Informatics Department, J. Craig Venter Institute, Rockville, Maryland, United States
{RMadupu,SYooseph,GSutton}@jcvi.org

Abstract. We aim to design a system for classifying scientific articles based on the presence of protein characterization experiments, intending to aid the curators populating JCVI's Characterized Protein (CHAR) Database of experimentally characterized proteins. We trained two classifiers using small datasets labeled by CHAR curators, and another classifier based on a much larger dataset using annotations from public databases. Performance varied greatly, in ways we did not anticipate. We describe the datasets, the classification method, and discuss the unexpected results.

Keywords: Classification, Biomedical Text Mining, Text Categorization, Database Curation, Imbalanced and Sparse Data.

1 Introduction

The Characterized Protein Database (CHAR) is a resource currently being developed by the J. Craig Venter Institute (JCVI) in support of their prokaryotic genome annotation pipeline, as well as broader annotation efforts. Records in the database include: protein name, gene symbol, organism name, GO terms, and synonymous accessions in other public databases. Moreover, each protein is linked to the scientific publications reporting the experimental results that characterize it. Taxon specific, functional information is drawn manually from the referenced publications by curators. CHAR is used within JCVI's auto-annotation pipeline to annotate novel prokaryotic gene products that are homologous to proteins already curated in CHAR. When CHAR is complete and populated, it is planned to be a high-quality publicly available resource for both prokaryotic and eukaryotic proteins.

As noted before by shared annotation tasks (e.g. KDD'02 [1], TREC Genomics 2004 [2], BioCreAtIvE [3]), the manual curation of each article is a slow process; the curator must locate articles about the species or protein of interest, and determine, by reading at least the abstract, if the article contains an experimental characterization. Our task is to automate parts of this process to reduce the amount of curator time

C. Blaschke and H. Shatkay (Eds.): ISBM/ECCB 2009, LNBI 6004, pp. 33–42, 2010.
© Springer-Verlag Berlin Heidelberg 2010

needed to populate CHAR. The goal is to develop a system for classifying journal articles as either *relevant* or *irrelevant* to CHAR, based on their title and abstract. Abstract text is used because it is readily available from the PubMed database [4]. Notably, about 84% of the 834 articles manually curated for this task were classifiable as either *relevant* or *irrelevant* to CHAR, based solely on their titles and abstracts.

To train classifiers, we created three datasets, each containing both *relevant* and *irrelevant* articles. Two small sets were built by manual curation and a much larger third set by using references to articles in public databases (Swiss-Prot and GO [5,6]). An additional validation set, consisting of *relevant* articles only, was formed from existing CHAR references that were curated before we began the task.

We use a multi-variate Bernoulli model [7] based on stemmed terms to represent each article, and train a standard naïve Bayes classifier. We expected the classifier trained on the large dataset to perform at least as well as the classifiers trained on the smaller sets. However, the opposite occurred. The latter classifiers outperformed the former on both the training and the held-out validation sets. Moreover, the classifiers trained on the small datasets come close to meeting the 70% recall and 80% precision requirements specified as useful by CHAR curators. This is despite the fact that they were trained only on a relatively small number of manually selected abstracts.

As the use of a large number of GO-curated articles for training a classifier was expected to boost performance, we attempt to explain the reduction in performance that actually occurred when using them. Our analysis shows that term distributions within the articles obtained from GO differ significantly from those associated with the other relevant datasets. Our experiments and results are described throughout the rest of the paper.

2 Dataset Construction

We built and used three datasets of abstracts taken from journal articles for training and testing, and an additional set for validation. The text for all abstracts was retrieved from PubMed [4]. The three training and test sets contain both *positive* examples, i.e. abstracts of articles *relevant* to CHAR, and *negative* (*irrelevant*) ones. Two of these datasets are relatively small, containing at most a few hundreds of manually curated abstracts, and were originally intended for validation only. The third set contains thousands of abstracts referenced from reliable public databases, and was originally intended for the training and testing process. The fourth dataset, used for independent validation, consists only of *relevant* articles (*positive* examples) that were already in CHAR at the onset of the project.

Our first dataset, referred to as the **Curated Journal** dataset, contains 96 *positive* and 107 *negative* abstracts. These were collected by a CHAR curator (RM) from every article in four issues from three different journals[1]. These journals are the ones most commonly referenced in CHAR, and three of the four respective issues each contributes at least one reference to CHAR. RM read the title and abstract of each article, and labeled it as either *relevant*, *irrelevant*, or, if unable to determine the relevance from

---

[1] J. of Bacteriology Vol. 189, #5, 2007, J. of Bacteriology Vol. 189, #15, 2007, J. of Biological Chemistry vol. 257, #19, 1982 and Molecular Microbiology vol. 33, #2, 1999.

the abstract alone, assigned it the label *maybe*. Articles labeled *maybe* were discarded from the dataset and are not used in our experiments. Notably, 86.8% of the examined articles were classifiable as *relevant or irrelevant* based on their title and abstract alone. Table 1 shows the dataset statistics.

Our second dataset, called the **Curated Swiss-Prot** dataset, consists of 324 *positive* and 174 *negative* abstracts. To build it, first 300 articles were selected at random from references within Swiss-Prot entries [5], and 300 were collected from entries in CHAR originally populated by an automated process which used Swiss-Prot information considered to suggest experimental characterization. As the reliability of these references was uncertain, RM manually labeled these documents, producing a dataset of 498 abstracts labeled with confidence as *relevant* or *irrelevant*. The statistics for the dataset are shown in Table 2. Again, articles labeled as *maybe* were discarded and are not used in our experiments. Of the 600 articles, 83% (498) were classifiable based on their titles and abstracts alone.

Table 1. Curation Labels for the **Curated Journal** Dataset. Numbers shown in boldface indicate documents actually used as *positive/negative* examples in our dataset.

	Relevant	*Irrelevant*	*Maybe*	Total
J. Bacteriol	71	40	20	131
J. Biol Chem	13	65	6	84
Mol Microbiol	12	2	5	19
Total	**96**	**107**	31	234

Table 2. Curation labels for the Curated Swiss-Prot Set. Numbers shown in boldface indicate documents actually used as positive/negative examples in the dataset.

Swiss-Prot articles	*Relevant*	*Irrelevant*	*Maybe*	Total
At random	85	155	60	300
From CHAR	239	19	42	300
Total	**324**	**174**	102	600

As the above datasets are fairly small, we planned to use them for validation only. For robust testing and training of a classifier, we aimed to build a much larger dataset, denoted **SP-GO**, utilizing curated labels assigned to journal articles by online public databases (Swiss-Prot [5] and GO [6]). We expected that adding many reliable, publically available, curated relevant examples would improve the classifier, supporting future automated curation in CHAR.

Swiss-Prot [5] entries hold references to PubMed articles, labeled to describe the information provided by the article. Articles labeled 'CHARACTERIZATION' satisfy the formal criterion of relevance to CHAR [8]. We thus collected all the articles in Swiss-Prot that were so labeled, providing 1,451 *positive* examples. The Gene Ontology project [6] also includes references to articles from PubMed that support ontology assignments. Evidence codes denote the type of information present in the referenced article. Articles assigned an experimental evidence code (EXP, IDA, IPI, IMP, IGI or IEP), satisfy the criterion of relevance to CHAR [9]. CHAR has even established an automated process for migrating such GO annotations into CHAR; as such, articles bearing the above GO codes suggest a reliable extensive data source. We thus collected all articles bearing the above evidence codes from GO, adding 8,403 *positive* examples.

Negative examples are harder to define: articles with non-experimental evidence codes in GO are not necessarily *irrelevant*, while articles in Swiss-Prot that are not labeled 'CHARACTERIZATION' may still carry experimental characterization [8,9]. To overcome this difficulty, and to obtain as *negative* examples articles associated with proteins that are unlikely to be experimentally characterized, we utilized flags attached to GenBank [10] sequence entries. GenBank entries typically carry an '*experimental*' flag for genes/proteins that have been experimentally characterized. CHAR curators estimate (based on experience) that in 80-90% of the cases where a flag is not present, the sequence is indeed not experimentally characterized. We thus gathered 10,012 GenBank entries not bearing the '*experimental*' flag, mapped the sequences (by identity) to their respective Swiss-Prot entries, and gathered all the articles referenced from these Swiss-Prot entries, resulting in 67,892 articles. Of these, 5.5% were found in the *positive* dataset, (which agrees with the estimate that only 80-90% of the un-flagged sequences are truly uncharacterized), and discarded. To have an equal number of positive and negative examples we selected 9,854 articles, at random, from the resulting negative pool.

The third dataset, **SP-GO**, thus consists of 9,854 *relevant* abstracts (1,451 Swiss-Prot, 8,403 GO), and 9,854 *irrelevant* abstracts, selected at random from the large negative set, as described above. As many of the articles in the *negative* set have an earlier publication date than many of the *positive* ones (data not shown), to avoid temporal artifacts in the classification (*conceptual drift* as noted in TREC Genomics [2]), the sampling of the 9,854 negative documents was biased toward recent publications.

Finally, as an independent validation set, we used all articles (255 abstracts) that were referenced from fully curated CHAR entries, as of May 2008, when we started the project. We refer to this dataset as the **CHAR** dataset. We note that this dataset consists of *positive* (*relevant*) abstracts only.

3 Methods and Classification

The titles and abstracts of the articles in all the datasets were downloaded from PubMed, tokenized into unigrams and bigrams, and stemmed using Porter stemming [11], to obtain a set of statistical terms. Stop words were removed, as were frequent terms (occurring in more than 60% of the abstracts) and rare terms, (occurring in fewer than 3 abstracts).

For classification, documents are represented using the multi-variate Bernoulli model [7], and a standard naïve Bayes classifier was implemented [12]. Under the naïve Bayes model the probability of a document given a class is calculated as:

$$\Pr(doc \mid class) = \prod_{\substack{terms\ in \\ document}} \Pr(term \mid class) * \prod_{\substack{terms\ not\ in \\ document}} (1 - \Pr(term \mid class)) .$$

We use a naïve Bayes classifier as it is simple to implement and modify, while its performance is comparable to that of other classifiers. The potential performance gain by using a different classifier (e.g. SVM [13,14]), is negligible compared to the differences in performance observed by varying the training sets as shown below.

Classification results are evaluated using recall, precision and accuracy, which are all defined in terms of number of *true positives* (*TP*), number of *true negatives* (*TN*), number of *false positives* (*FP*) and *number of false negatives* (*FN*), as:

$$recall = \frac{TP}{TP+FN}, \quad precision = \frac{TP}{TP+FP}, \quad accuracy = \frac{TP+TN}{TP+FP+TN+FN}.$$

To examine the differences between the various datasets, we calculate the Kullback-Leibler (KL) divergence [15], among the term distributions within the datasets, where the distribution per dataset is a multinomial over all the terms used to represent the abstracts. The KL divergence measures the difference between two probability distributions. We use a symmetric version [15] defined, for two multinomial probability distributions over n events, P and Q where $P = \{p_1,...,p_n\}$ and $Q = \{q_1,...,q_n\}$, as:

$$KL(P,Q) = \sum_{i=1}^{n} (p_i - q_i) \log_2 \frac{p_i}{q_i}.$$

The KL divergence is non-negative; the lower it is, the more similar the distributions are.

4   Experiments and Results

According to CHAR curators, for a classifier to be useful for their task, its recall should be at least 70% and its precision at least 80%. With this in mind, we initially trained a naïve Bayes classifier using the *SP-GO* dataset. Given the magnitude and the design of the dataset (where abstracts were chosen based on curator input and annotations in public databases), we expected a classifier trained on it to perform well. We tested the classifier on both the *Curated Swiss-Prot* and *Curated Journal* datasets. Contrary to our expectations, the *SP-GO* trained classifier performs poorly, placing an incorrect label on more than half of our manually curated articles, as is reported in Table 3.

Table 3. *SP-GO* performance on the *Curated Journal* and *Curated Swiss-Prot* datasets

	Curated Journal	*Curated Swiss-Prot*
Recall	30.21%	14.20%
Precision	34.52%	51.11%
Accuracy	39.90%	35.34%

To ensure that, in each dataset, the *positive* vs. *negative* abstracts are indeed distinguishable and "classifiable", we performed complete 5-fold cross-validation, repeated 10 times (using 10 different 5-fold-splits), for each of the three datasets. Table 4 shows the results, averaged over the 5-folds and over all the 10 complete cross-validation runs.

The classifiers based on each set all perform well in the cross-validation setting. All satisfy the requirement of 70% recall, and the **Curated Swiss-Prot** set satisfies 80% precision as well. These results clearly show that within each dataset, classification is possible.

To verify that a classifier trained on one dataset can still perform well on another, we trained a classifier using each of the small datasets (**Curated Journal** and **Curated Swiss-Prot**), and tested on the other. Table 5 shows the results. The performance is much better than that of the classifier based on **SP-GO,** despite the smaller training sets, which were each drawn from a different data source.

For a final validation, we used each of the three trained classifiers (the above two and the one trained on **SP-GO**) to classify the **CHAR** dataset, which consists of references that were already in CHAR when our project began. Results are shown in Table 6. As the **CHAR** dataset contains only *relevant* (*positive*) articles, only accuracy can be reported.

While the classifier trained on the **SP-GO** dataset labels most of the **CHAR** abstracts as irrelevant, the other two classifiers do label most of the articles correctly, even though they were trained on relatively little data. This is despite the fact that articles in **SP-GO** were curated and labeled in public databases using criteria seemingly equivalent to those employed by CHAR.

Table 4. Average results from 10 times 5-fold cross-validation classification performed over each dataset. (Standard deviation in parentheses.)

	SP-GO	Curated Journal	Curated Swiss-Prot
Recall	73.3% (0.12%)	72.5% (2.92%)	89.4% (0.77%)
Precision	73.4% (0.14%)	71.0% (1.70%)	84.2% (0.53%)
Accuracy	73.4% (0.12%)	73.0% (1.85%)	82.2% (0.37%)

Table 5. Performance of classifiers trained over one curated set and tested on the other

Trained with:	Curated Journal	Curated Swiss-Prot
Tested on:	Curated Swiss-Prot	Curated Journal
Recall	80.86%	93.75%
Precision	72.98%	52.33%
Accuracy	68.07%	56.65%

Table 6. Accuracy of the three classifiers, over the **CHAR** dataset

	SP-GO	Curated Journal	Curated Swiss-Prot
Accuracy	26.67%	80.00%	87.84%

To further explore the difference between the **SP-GO** and the other datasets, we calculate the Kullback-Leibler (KL) divergence between the term distributions associated with each dataset, separately examining the *positive* articles and the *negative* ones in each dataset. Ideally, the term distributions of two *positive* sets should be similar (low divergence) while term distributions between any pair of *positive* and *negative* sets should show much difference (higher divergence).

Table 7 shows the KL divergence between pairs of the *negative/positive* classes from all the datasets as well as the **CHAR** set. These values show that the *positive* abstracts in the small sets (*Curated Swiss-Prot*, *Curated Journal*) and in *CHAR* share relatively similar term distributions (relatively low KL divergence), while the difference between every pair of their *positive* and *negative* subsets is much higher in terms of KL divergence. In contrast, both the *negative* and the *positive* subsets of *SP-GO* are almost equally dissimilar to the *positive* subset of *Curated Journal*; moreover, the *positive* subset of *SP-GO* is less similar to both the *CHAR* dataset and to the *positive* subset of *Curated Swiss-Prot* (higher KL), than the *negative* *SP-GO* subset is (see italicized numbers in Table 7). These properties of the *SP-GO* term distributions, are consistent with the results reported above, that the classifier trained on *SP-GO* labeled most of the *CHAR* articles as irrelevant (Table 6).

As the *SP-GO* *relevant* (*positive*) abstracts originated from two distinct sources, namely Swiss-Prot and GO, to better understand the phenomenon, we separated *SP-GO* into two: the 1,451 abstracts collected from Swiss-Prot (denoted *SPonly*) and the 8,403 abstracts collected from GO (*GOonly*). Table 8 shows the KL divergence between the term distributions of these two sets and the other datasets.

Table 7. Kullback-Leibler Divergence between dataset pairs (*CurSP* denotes our *Curated Swiss-Prot* dataset; *CurJol* denotes our *Curated Journal* dataset. *Pos* indicates the *positive* portion of the dataset, *Neg* indicates *negative*.)

	CurJol Pos	*CurJol* Neg	*CurSP* Pos	*CurSP* Neg	*CHAR* (Pos)
SP-GO Pos	1.1442	1.2714	0.9622	0.5624	0.9224
SP-GO Neg	1.1733	1.3784	0.5790	0.2200	0.6752
CurJol Pos			0.8167	1.5319	0.7302
CurJol Neg			1.4791	1.7227	1.3580
CurSP Pos					0.3175
CurSP Neg					0.9667

Table 8. KL divergence between each of the *SPonly* and *GOonly* sets and all other subsets. (*CurSP* denotes our *Curated Swiss-Prot* dataset; *CurJol* denotes our *Curated Journal* dataset. *Pos* indicates the *positive* portion of the dataset, *Neg* indicates *negative*.)

	CurJol Pos	*CurJol* Neg	*CurSP* Pos	*CurSP* Neg	*CHAR* (Pos)
SPonly (*Pos*)	0.9541	1.2336	0.3825	0.5031	0.5010
GOonly (*Pos*)	1.3226	1.3920	1.2130	0.6744	1.1381

Table 8 clearly shows that abstracts labeled as bearing experimental characterization by Swiss-Prot (*SPonly*) are more similar in their term-distribution to the *positive* abstracts in our hand-curated sets than to the *negative* abstracts. In contrast, term-distributions of the abstracts that contain experimental characterization based on GO (*GOonly*) are dissimilar to the distributions of the *positive* abstracts in our curated sets, and actually even more similar to the term distributions of the *negative* abstracts in one of the sets (*Curated SwissProt*). The apparent discrepancy between GO's notion of abstracts bearing experimental characterization and CHAR's is particularly noteworthy, given that GO curated abstracts were considered as a mainpossible source in support of CHAR's curation.

A naïve Bayes classifier trained on the 1,451 *SPonly positives* and 1,451 *negative* articles selected at random from the *SP-GO negatives*, demonstrates improved performance, closer to that obtained by classifiers trained on the small curated datasets. Performance is best on the **Curated Swiss-Prot** dataset where both the recall and the precision surpass the requirements set out by the CHAR curators. The results are shown in Table 9.

Table 9. SPonly performance on the **Curated Journal**, **Curated Swiss-Prot** and **CHAR** dataset

	Curated Journal	**Curated Swiss-Prot**	**CHAR** (Pos)
Recall	73.96%	70.37%	
Precision	52.21%	83.82%	
Accuracy	55.67%	71.89%	70.98%

When attempting to explain the results, our initial hypothesis was that terms indicative of a specific species or of species from specific kingdoms (such as common names, scientific names or strain identifiers) may be responsible for the differences in term distributions between the datasets. We found that terms typically associated with common model organisms, such as *E.coli*, mouse and yeast, are overrepresented in *relevant* articles, as proteins from these species are more likely to be experimentally characterized; we also found that articles from Swiss-Prot and from GO report on species from different kingdoms at different rates[2]. We have conducted experiments, completely removing all terms indicative of a specific species or, alternatively, replacing such terms with the species-generic pseudo-term SPECIES_PLACEHOLDER to capture all occurrences of a species-related term in the abstracts. While these strategies improved results slightly (1-2%), and may help prevent bias for certain species in future classification, this small effect does not explain the major difference in performance across the different datasets.

Our current hypothesis is that articles from GO cover a few types of experiments that are relevant to CHAR heavily and rarely mention others. The articles we collected from GO are not evenly distributed across the six experimental evidence codes (EXP, IDA, IPI, IMP, IGI or IEP). Approximately 70% of the articles are associated with either 'Inferred from Direct Assay' (IDA) or 'Inferred from Physical Interaction' (IPI). These codes are both strongly related to binding assay experiments [9]. Related terms such as **interact** or **complex** are therefore highly overrepresented in the **SP-GO** positive set, which mostly consists of articles from GO. It is likely that the **SP-GO** positives contain many articles based only on a few forms of experimental characterizations, leading to a classifier that labels articles reporting other types of experiments as *irrelevant*, regardless of their actual relevance to CHAR This may contribute to the low recall shown by the **SP-GO** classifier in Table 3, and may explain why classification performance is improved by removing the GO articles.

The *negative* articles in the **SP-GO** dataset were collected from Swiss-Prot entries of proteins that, based on the selection process, are expected to be uncharacterized. Such articles, which are associated with protein sequences that have not been experimentally

---

[2] PubMed references in GO (which make up 85% of the **SP-GO** *positives*) were 67% prokaryote, 14% eukaryote and 20% virus when we collected them. Sequence entries in Swiss-Prot when we collected our articles (100% of the **SP-GO** *negatives*) were 57% prokaryote, 36% eukaryote, 4% archaea and 3% virus [16].

characterized, are likely to be initial papers pertaining to the proteins, reporting their genomic and proteomic sequence. This is corroborated by the fact that the terms **sequence**, **genome** and **cdna** are highly overrepresented in the *SP-GO negatives*. It is likely that the *SP-GO negative* set thus contain many *irrelevant* articles discussing sequences, while it lacks *irrelevant* articles of other types. Furthermore, since the *negative* articles in the **Curated Journal** dataset are less likely to discuss sequence data than the *negative* articles in the *SPonly* dataset, classifiers trained on the *SPonly* dataset may erroneously label the *irrelevant*, non-sequence articles in the **Curated Journal** dataset as *positive*, which helps explain the low precision and low overall accuracy of the *SPonly* classifier over the **Curated Journal** dataset shown in Table 9.

5 Conclusion

We trained naïve Bayes classifiers to identify abstracts that are likely to describe experimental characterization of proteins (*relevant*), as opposed to abstracts unlikely to contain such information (*irrelevant*). The classifiers are intended to support the curation of abstracts for JCVI's CHAR (Characterized Protein) database. To train and test the classifiers we constructed small hand-curated datasets, as well as a large set based on previously curated abstracts from Swiss-Prot and GO. We expected the latter set to support training a well-performing classifier, given the dataset size and the careful consideration invested in its construction. Specifically, given that, by definition, articles containing experimental characterization are relevant to CHAR.

While the abstracts bearing the 'CHARACTERIZATION' label in Swiss-Prot proved to be effective *positive* examples, articles bearing experimental characterization flags in GO did not. The latter was particularly surprising, as GO was considered up to this point as the most likely source of information for CHAR.

Most notably, and on the positive side, we have shown that classifiers trained on relatively small hand-curated datasets perform at a high level – very close to the level required by CHAR curators to be useful – both when classifying the other sets' documents, and when classifying the left-out validation set (CHAR).

Another interesting finding of this study was that in more than 80% of the manually examined articles, the abstract and title alone contained sufficient information for determining the relevance of the article for the CHAR curation. While the actual evidence is most likely to be found in the full text, it is important to note that the coarser task of just determining relevance can be performed (in most cases) using the title and the abstract alone.

References

1. Yeh, A., Hirschman, L., Morgan, A.: Background and Overview for KDD Cup 2002 Task 1: Information Extraction from Biomedical Articles. In: ACM SIGKDD Explorations Newsletter (2002)
2. Cohen, A., Bhupatiraju, R.T., Hersh, W.R.: Feature Generation, Feature Selection, Classifiers, and Conceptual Drift for Biomedical Document Triage. In: 13th Text Retrieval Conference -TREC 2004, Gaithersburg, MD (2004)

42   R. Denroche et al.

3. Blaschke, C., Leon, E.A., Krallinger, M., Valencia, A.: Evaluation of BioCreAtIvE assessment of task 2. BMC Bioinformatics 6 (Suppl. 1), S16 (2005)
4. PubMed, http://www.ncbi.nlm.nih.gov/pubmed
5. Swiss-Prot Protein Knowledgebase, http://ca.expasy.org/sprot/
6. The Gene Ontology Project, http://www.geneontology.org/
7. McCallum, A., Nigam, K.: A Comparison of Event Models for Naive Bayes Text Classification. In: Learning for Text Categorization Workshop, AAAI 1998 (Tech. Report WS-98-05) (1998)
8. Swiss-Prot Protein Knowledgebase: A Primer on UniProtKB/Swiss-Prot Annotation, http://www.uniprot.org/docs/annbioch
9. The Gene Ontology Project: Guide to GO Evidence Codes, http://www.geneontology.org/GO.evidence.shtml
10. GenBank, http://www.ncbi.nlm.nih.gov/Genbank
11. Porter, M.F.: An Algorithm for Suffix Stripping. Program 14, 130–137 (1980)
12. Duda, R.O., Hart, P.E., Stork, D.G.: Pattern Classification, 2nd edn. John Wiley & Sons, Inc., New York (2001)
13. Joachims, T.: Text Categorization with Support Vector Machines: Learning with Many Relevant Features. In: Nédellec, C., Rouveirol, C. (eds.) ECML 1998. LNCS, vol. 1398. Springer, Heidelberg (1998)
14. Donaldson, I., Martin, J., de Bruijn, B., Wolting, C., Lay, V., Tuekam, B., Zhang, S., Baskin, B., Bader, G.D., Michalickova, K., Pawson, T., Hogue, C.W.V.: PreBIND and Textomy – mining the biomedical literature for protein-protein interactions using a support vector machine. BMC Bioinformatics 4(11) (2003)
15. Kullback, S., Leibler, R.A.: On Information and Sufficiency. Annals of Mathematical Statistics 22, 79–86 (1951)
16. Swiss-Prot Protein Knowledgebase: Release Notes for UniProtKB Release (July 22, 2008), http://www.expasy.ch/txt/old-rel/relnotes.56.htm

Mining Protein-Protein Interactions from GeneRIFs with OpenDMAP

Andrew D. Fox[1,*], William A. Baumgartner Jr.[2], Helen L. Johnson[2],
Lawrence E. Hunter[2], and Donna K. Slonim[1,3,*]

[1] Department of Computer Science, Tufts University, Medford, MA 02155
[2] Center for Computational Pharmacology,
University of Colorado School of Medicine, Aurora, CO 80045
[3] Department of Pathology, Tufts University School of Medicine, Boston, MA 02111
{Andrew.Fox,Donna.Slonim}@tufts.edu

Abstract. We applied the OpenDMAP [1] and BioNLP-UIMA [2] NLP
systems to the task of mining protein-protein interactions (PPIs) from
GeneRIFs. Our goal was to assess and improve system performance on
GeneRIF text. We identified several classes of errors in the system's output on a training dataset (most notably difficulty recognizing protein
complexes) and modified the system to improve performance based on
these observations. To improve recognition of protein complex interactions, we implemented a new protein-complex-resolution UIMA component. We added a custom entity identification engine that uses GeneRIF
metadata to annotate proteins that may have been missed by the other
engines. These changes simultaneously improved both recall and precision, resulting in an overall improvement in F-measure (from 0.23 to
0.48). Results confirm that the targeted enhancements described here
lead to a substantial improvement in performance.

Availability: Annotated data sets and source code for the new UIMA
components can be found at http://bcb.cs.tufts.edu/GeneRIFs/

1 Introduction

Without sufficiently accurate automated tools for the extraction of protein-
protein interactions (PPIs) from natural language text, the task is left to human
PPI database curators. However, with the overwhelming number of new biomedical journal articles each year, the case for manual curation is losing ground and
the need for viable alternatives is rapidly increasing [3]. Software systems are or-
ders of magnitude faster than human curators but currently exhibit much lower
prediction accuracy [4]. Hence, there is value in designing high-throughput software filters that can recommend sentence fragments to a human curator for
further review. Even in a recommendation role, such systems should be as accurate as possible, although there may be significant differences between individual
curators in their preferences for high-recall or high-precision systems [5].

---

* Corresponding author.

C. Blaschke and H. Shatkay (Eds.): ISBM/ECCB 2009, LNBI 6004, pp. 43–52, 2010.

The aim of our work is to assess and improve the performance of a state-of-the-art PPI recognition system using OpenDMAP [1] on a restricted target domain, namely GeneRIFs. System modifications are made based on the system's performance on a training data set, and the performance of the modified system is then compared to that of the original system on an independent test set.

GeneRIFs are short (less than 256-character) text summaries of gene function, based on and linked to primary publications about the supporting data. They are submitted both by National Library of Medicine staff scientists and by other members of the scientific community. Not all GeneRIFs are comprised of complete sentences, and typographical or grammatical errors can remain in the database for some time until reviewed and corrected (either manually or by automated systems [6]).

However, GeneRIF text is an increasingly abundant source of information summarizing crucial functional roles of genes and gene products. Their value and impact are reflected by the rapid growth in the size and scope of the database: from 27,645 GeneRIFs describing 9,126 genes in eight model organisms in 2003 [7], to 157,280 GeneRIFs describing 29,297 distinct Entrez Gene entries in 571 organisms in 2006 [6], to 270,461 GeneRIFs describing 48,931 Entrez Gene entries in 1,195 organisms as of April 21, 2009. While this growth rate still appears insufficient to obviate the need for automated curation [3], GeneRIFs are becoming an increasingly valuable annotation resource. Furthermore, methods for computationally extracting GeneRIFs have shown promise [8] and potentially provide a mechanism for rapidly raising the level of coverage.

We chose to focus on the GeneRIF target domain because we expected that the restricted length of GeneRIFs would lend itself to the potential for enhanced PPI recognition [9,10]. Additionally, all GeneRIFs must be linked to at least one valid Entrez Gene identifier. These links, collectively described here as GeneRIF "metadata," can potentially provide an additional mechanism for entity identification. Early work on mining protein-protein interactions demonstrated that delegating the task of entity identification to humans offers a huge advantage [11]. While recent work has shown that automated systems are improving substantially [12], human annotation remains the gold standard. With GeneRIFs, we can partially delegate this task to humans simply by relying on the GeneRIF curators' links between the text and specific genes.

OpenDMAP is an ontology-driven, rule-based concept recognition system that uses patterns to extract concepts and relations among concepts from text. It was one of the overall best performers on the BioCreative-II Interaction Pair Subtask [4], with 31% averaged recall and 39% averaged precision on their full test data set, resulting in an averaged F-measure of 0.29. For the work described here, we first ran a variant of this system on our GeneRIF training data.

Examining the results, we noticed that despite the presence of three entity identification engines in the pipeline, many of the errors were caused by incorrect entity identification, particularly in the case of protein complexes. For example, consider the following GeneRIF: "Grh1-Bug1 complex contributes to a redundant network of interactions that mediates consumption of COPII vesicles"(PubMed ID: 17261844). The original system identifies "Grh1-Bug1" as a single token, and therefore OpenDMAP considers it a single potential interactor. Thus, the system is unable to detect the complex between Grh1 and Bug1, even though the phrase would be expected to match a simple protein-complex interaction pattern of the form: "[interactor1]-[interactor2] complex."

Another problem can be seen from examining this GeneRIF: "an alternative product of the mastermind locus that represses notch signaling" (PubMed ID: 12049771). The core entity identification engines in our system fail to identify "notch" as a protein. However, this GeneRIF is linked through the NCBI metadata to Drosophila *notch* (Entrez Gene ID: 31293). A tagger that explicitly looks for known aliases for that gene (which also include the English words "chopped," "facet," "split," and "strawberry") might recognize *notch* as a potential interactor. We expected that any recall gained by such rules would involve little or no loss of precision, because the chance that a GeneRIF linked to the *notch* gene would also include, say, an unrelated use of the word "strawberry" is negligible.

2   Methods

2.1   Training and Test Set Construction

To evaluate system performance, we developed manually-annotated training and test data sets of protein-protein interactions in GeneRIFs. We started by randomly selecting GeneRIFs from *S. cerevisiae*, *D. melanogaster* and *M. musculus* with metadata linked to two or more Entrez gene entries. These "multiply-linked" GeneRIFs, some of which presumably described interactions between some of the linked genes, made up approximately 75% of the data set. The remaining entries were randomly chosen GeneRIFs from the same species linked to only a single gene, and thus supposedly not describing interactions[1].

We divided the annotated data evenly into training and test data sets. Two scientists independently manually annotated the protein-protein interactions in the full training and test sets. All gene name mentions within each GeneRIF were identified, and all possible pairs of those gene names were evaluated by the annotators as positive or negative examples of protein-protein interactions. Positive examples included only those cases in which the two proteins were part of the same protein complex, or had a direct or indirect protein-protein interaction between them. Any case in which the text did not specifically describe an interaction was considered a negative example. Inter-annotator agreement was above 90% on both the training and test data sets from all three species. Discrepancies were then discussed by both annotators and nearly all were resolved. A handful of the GeneRIF sentences (3 in yeast and 1 in fly) were found to be genuinely ambiguous and were removed from the data set entirely. Table 1 summarizes the resulting numbers of GeneRIFs for the training and test data sets in each of the three species.

---

[1] Because the metadata linking GeneRIFs to Entrez Gene entries are not always complete, we observed a few instances where these singly-linked GeneRIFs actually did describe interactions.

Table 1. Training set and test set summary statistics. This table lists the number of GeneRIFs and the total number of gene-name pairs found in each data set.

	# GeneRIFs	# Gene Pairs
Mouse Train	35	96
Mouse Test	37	90
Fly Train	67	116
Fly Test	69	131
Yeast Train	111	197
Yeast Test	108	162
TOTAL	425	789

2.2   Overview of System Architecture

The original system was constructed as an Unstructured Information Management Architecture (UIMA) [13] pipeline. Sentence detection, tokenization and part-of-speech tagging are handled by the open-source LingPipe [14] NLP library. The system utilizes three commonly-used named entity identification (EI) engines: LingPipe [14] and two instances of Abner [15] trained on corpora from the NLPBA and BioCreative I challenges, respectively.

The system uses a 'union' model of protein tag aggregation, in which the union of all tags from the three EI engines is taken as the final set of protein tags. This aggregation model introduces the issue of overlapping protein tags, so these are resolved by an overlapping-annotation-resolution UIMA component. In the original system, this component resolved overlapping protein tags by removing any tag subsumed by a longer one, i.e. in a 'longest-first' approach. In our enhanced system, we resolve overlaps in a 'shortest-first' manner, so that protein tags meet the requirements of the protein complex resolver described below.

Finally, the OpenDMAP concept recognition component matches the GeneRIF text against a set of user-defined protein-protein interaction 'patterns' and reports all matched interactions. OpenDMAP allows the user to specify these concept recognition patterns using a powerful built-in pattern definition language that exploits ontological constraints to reduce ambiguity and allows the mixing of syntactic and semantic elements [1]. A prototypical OpenDMAP pattern for this task would look like: "[protA] {int-verb} {prep} {article}? {species}? [protB]", where [protA] and [protB] are the interacting proteins, {int-verb} indicates a relevant interaction verb, {prep} indicates a relevant interaction preposition, and '?' denotes optional elements. Relevant interaction verbs, nouns and prepositions are specified in the pattern files.

2.3   Metadata Protein Tagger

Each GeneRIF has accompanying metadata that lists the Entrez Gene IDs to which it refers. We added an additional entity tagging engine (the 'MetaTagger') that searches for and tags all protein name mentions corresponding to a known alias of a metadata-linked gene. As with the other EI engines, this tagging method results in an imperfect entity tagger, as the metadata may be incomplete or incorrect, or the author may have referred to the linked protein using a non-standard alias that the system cannot recognize.

As we employ the 'union' model of tag aggregation, it is beneficial to incorporate additional high-precision tagging engines (such as the MetaTagger) into the processing pipeline. This provides a good chance of improving recall without lowering precision significantly. In fact, in our system the MetaTagger increases precision as well as recall, because the curated metadata allows it to tag some protein name mentions more precisely than the other EI engines, reducing the number of false positives. The following GeneRIF is one where the MetaTagger increases recall: "Dysfusion dimerizes with the Tango bHLH-PAS protein..." (PubMed ID: 17652079). None of the three predictive protein tagging engines in the system's UIMA pipeline identify "Dysfusion" at all and together they identify 'Tango bHLH-PAS protein' instead of simply 'Tango.' Using the Entrez Gene metadata, the MetaTagger engine finds and tags both 'Dysfusion' and 'Tango' correctly, allowing OpenDMAP to identify their dimerization.

2.4 Protein Complex Resolver

To improve protein complex recognition, we created a new protein-complex-resolution UIMA component and added it into the UIMA pipeline after the overlapping annotation resolver. The new component classifies each token as either describing a protein complex or not, based on a set of straightforward heuristics (described below) that are designed to recognize binary and ternary complexes that are identified by the names of their components. Each token classified as a protein complex is then split into its multiple constituent tokens, and these tokens are tagged as proteins where appropriate. For example, a token such as 'Sir3-Sir4' would be split into three tokens: 'Sir3', '-' and 'Sir4', and both 'Sir3' and 'Sir4' would be tagged as proteins if they had not previously been tagged as such. This token-splitting is necessary because OpenDMAP's concept recognition engine is not designed to match interaction concepts that are expressed within a single token. We note that it is undesirable to tokenize on '-' and '/' characters universally, as this can introduce errors at the entity identification stage that propagate to all later stages in the pipeline.

The protein complex resolver identifies protein complex tokens to be split by recognizing two distinct cases. The first case occurs when there is a protein-complex associated keyword (see Table 2) immediately to the left or right of the hyphenated token. In this case the protein complex resolver splits the appropriate token if at least $(n-1)$ of the n potentially interacting proteins in the supposed complex had protein tags generated for them by the protein tagging engines (after tag aggregation). Note that the MetaTagger is one of these tagging engines, so the metadata can help in recognizing potential protein complexes.

The second case occurs when no such keyword is present. In this case, the system requires that all of the n potentially interacting proteins have distinct protein entity tags associated with them. This heuristic scheme was observed

Table 2. List of protein-complex associated keywords used by the Protein Complex Resolver. Starred items must occur immediately after of the protein complex mention, e.g. "Sir3-Sir4 dimers." Non-starred items may occur either immediately before or immediately after the mention, e.g., "the complex Sir3-Sir4."

complex	heterodimer	dimer	trimer
complexes*	heterodimers*	dimers*	trimers*
heterotrimer	heteromer	interaction*	
heterotrimers*	heteromers*	interactions*	

in the system training phase to deal successfully with both binary complexes ($n = 2$) and ternary complexes ($n = 3$).

2.5   Enhancements to OpenDMAP Recognition Patterns

After assessing the errors made on the training data, we added, removed or modified interaction recognition patterns as necessary to correct as many of the observed errors as possible. Candidate changes were either accepted or rejected based on re-assessing classification error after each individual change. To avoid overfitting to the training dataset, we rejected changes that we expected would generalize poorly to unseen examples. For example, the pattern "[protA] {int-verb} {prep} _ and also {prep}? [protB]" improves recall on the training set, but requires the wildcard '_' element (which matches *any* string) to do this. The presence of the wildcard element is likely to cause this pattern to match a significant number of sentence fragments that do not actually specify an interaction between [protA] and [protB]. In general, we rejected candidate patterns that increased recall at unacceptable costs in precision on the training data.

We also added two new syntactic concept classes, 'adverb' and 'article', as well as a new semantic concept class 'species.' These concept classes proved useful in recognizing interactions that the original system overlooked because of the presence of an unexpected article or species modifier preceding a protein name, or an unexpected adverb in an interaction phrase.

3   Results

We evaluated the performance of the resulting system on the test data sets described above. In addition, we compared these results on the same test data to two other versions of the system: a variant of the original BioCreative II version of the OpenDMAP PPI-recognition system, and a version of our system without the MetaTagger (so that we could assess the impact of this step). The recall, precision, and F-measure for all of these test sets are shown in Table 3.

We first see that the difficulty of recognizing protein-protein interactions in text depends greatly on the species being discussed. This is not surprising, because entity identification is a major challenge for this task, and we deliberately chose species with a range of different gene naming conventions. Yeast genes

Table 3. Summary of PPI recognition results on test data sets. (orig) indicates results of the original OpenDMAP system; (-meta) indicates results on the modified system but without the metadata protein tagger; (+meta) indicates results on the fully modified system. "Combined" results aggregate results across all three species, normalized to account for different test set sizes between species.

	Recall	Precision	F-Measure
Mouse (orig)	0.14	0.56	0.22
Mouse (-meta)	0.38	0.52	0.44
Mouse (+meta)	0.38	0.56	0.45
Fly (orig)	0.07	0.40	0.12
Fly (-meta)	0.19	0.41	0.26
Fly (+meta)	0.31	0.47	0.38
Yeast (orig)	0.27	0.50	0.35
Yeast (-meta)	0.55	0.62	0.59
Yeast (+meta)	0.56	0.63	0.60
Combined(orig)	0.15	0.50	0.23
Combined(-meta)	0.37	0.54	0.44
Combined(+meta)	0.42	0.56	0.48

have relatively constrained gene symbols, typically consisting of three letters and a number (e.g., *PKC1*). Mouse gene names are more variable, and the fruit fly community seems to take particular delight in creating and using whimsical names comprised of common English words (e.g., *boss*, a.k.a. *bride of sevenless*). While this tradition makes for some amusing reading, it wreaks havoc with the automated entity identification process. Consistent with previous work on gene name normalization [12,16], therefore, our systems perform best on yeast, do moderately well on mouse, and struggle with much of the fly data. In order to avoid positively biasing the "combined" results reported in Table 3 (due to having more yeast data than fly or mouse data), the combined results are normalized so that each of the three species is weighted equally.

3.1 Gold Standard Entity Identification (GSEI) Analysis

Given the results, an interesting question to ask is how much of the remaining error is due to incorrect entity identification and how much is due to incorrect recognition of interaction relations between entities. To answer this question, we created a version of the modified system which used the set of gold-standard protein name tags directly instead of having to infer them computationally. This system was able to achieve an overall recall of 0.48, precision of 0.83 and F-measure of 0.61. Hence, for the modified (+meta) system, entity identification errors are affecting precision much more than recall (precision loss of 27 points vs. recall loss of only 6 points). We also compared the original system (orig) to its GSEI variant and found similar results: in the GSEI system, precision increased from 0.50 to 0.84 and recall increased from 0.15 to 0.21.

Clearly, gold standard entity identification greatly reduces the proportion of false positive interactions (from 44% of all predictions down to only 17%).

However even with perfect entity identification, recall increases to only 48%. The remaining 52% of lost recall is then due to a lack of coverage (by the pattern set) of the space of all possible language constructs that can be used to describe interactions. To try to address this issue, the pattern set could be enhanced based on further training against additional manually annotated GeneRIFs, or through comprehensive review by an expert PPI curator.

4   Discussion and Future Work

Recall and F-measure are improved substantially on all species by our proposed changes to the system. On the combined data, we have simultaneously improved recall from 15% to 42% and precision from 50% to 56%, resulting in an overall improvement in F-measure from 0.23 to 0.48. Furthermore, the results verify that adding the MetaTagger EI engine produced a small but noticeable contribution in overall performance (a 5 point improvement in recall and a 2 point improvement in precision).

We note that the original system's performance in this restricted GeneRIF domain is not identical to OpenDMAP's performance on the BioCreative II Interaction Pair Subtask. This is most likely due to a combination of factors such as differences in species distribution, sentence complexity and annotation criteria (e.g., indirect interactions are not included for BioCreative, but are included here). In addition, the BioCreative II task required an additional gene normalization component that was not present in the original system described here. Thus, the two sets of numbers are not directly comparable. Nonetheless, our results appear to support the hypothesis that GeneRIFs' restricted structure and metadata are valuable for the identification of protein-protein interactions.

Though automatically extracting PPIs from *manually*-curated GeneRIFs may seem to simply shift the annotation bottleneck, automatic mining of GeneRIFs is already possible [8]. The limitation of such GeneRIFs is that they lack metadata directly linking them to Entrez Gene entries. However, the articles from which they are mined *are* often linked to specific genes. Future work might therefore investigate the degree to which these links are useful as metadata for mining PPIs from computationally inferred GeneRIFs.

We also note that the metadata tagger improved recall dramatically in fly but had much lower impact for the yeast and mouse data. This reflects the facts that entity identification in fly is noticeably more difficult than in the other species [4] and that using human curated EI metadata can partially compensate for this difficulty. These results also suggest that most GeneRIFs in the yeast and mouse datasets use recognizable versions of gene names, or else that the list of aliases currently provided to the metadata tagger is not as valuable for these species.

From the GSEI results it is also clear that a key future challenge in protein-protein interaction extraction will be in recognizing the substantial fraction of interactions that are expressed in complex language structures and that may not be recognizable by currently existing systems. Even under an assumption of gold standard entity identification where precision is less of an issue, improving recall remains a significant challenge.

Improving recall on this task is not only important for a standalone system, but is also important in the assisted-curation setting because a high-recall recommendation system is likely to speed up curation by reducing the amount of unannotated text the curator must read in detail [17]. However, it remains to be seen what level of recall performance will be useful to a curator on this task in the assisted-curation setting [4].

4.1 Future Work

Future work will include extending the metadata tagger to incorporate additional gene designations, to identify English descriptions of complex gene names, and perhaps also to recognize imperfect matches to longer gene descriptions. Further improvement to entity identification may also be realized by incorporating an additional high-precision tagging engine [18].

Finally, improvements in PPI recognition may also be realized through some level of limited automation of the pattern training process. An attribution scheme could be built into the system to calculate recall and precision of individual patterns. A search and cross-validation approach could then be used to optimize the pattern set, provided that sufficient training data were available.

Acknowledgments. We would like to thank Kevin Cohen, Zhiyong Lu, and other members of the Hunter Lab and the Tufts BCB Group for helpful discussions and feedback. We gratefully acknowledge the support of the following grants from the National Library of Medicine: R21LM009411 (DKS, ADF, and LEH); R01LM009254 and R01LM008111 (WAB, HLJ, and LEH). The content is solely the responsibility of the authors and does not necessarily represent the official views of the National Library of Medicine or the National Institute of Health.

References

1. Hunter, L., Lu, Z., Firby, J., Baumgartner, W., Johnson, H., Ogren, P., Cohen, K.: OpenDMAP: An open source, ontology-driven concept analysis engine, with applications to capturing knowledge regarding protein transport, protein interactions and cell-type-specific gene expression. BMC Bioinformatics 9(1), 78 (2008)
2. BioNLP UIMA Component Repository, http://bionlp-uima.sourceforge.net/
3. Baumgartner, W.A., Cohen, K.B., Fox, L.M., Acquaah-Mensah, G., Hunter, L.: Manual curation is not sufficient for annotation of genomic databases. Bioinformatics 23(14) (2007)
4. Krallinger, M., Leitner, F., Rodriguez-Penagos, C., Valencia, A.: Overview of the protein-protein interaction annotation extraction task of BioCreative II. Genome Biol. 9(Suppl. 2), S4: 41–55 (2008)
5. Winnenburg, R., Wachter, T., Plake, C., Doms, A., Schroeder, M.: Facts from text: can text mining help to scale-up high-quality manual curation of gene products with ontologies? Brief Bioinform. 9(6), 466–478 (2008)
6. Lu, Z., Cohen, K.B., Hunter, L.E.: GeneRIF quality assurance as summary revision. In: Pac. Symp. Biocomput., pp. 269–280 (2007)

7. Mitchell, J.A., Aronson, A.R., Mork, J.G., Folk, L.C., Humphrey, S.M., Ward, J.M.: Gene indexing: characterization and analysis of NLM's GeneRIFs. In: AMIA Annu. Symp. Proc., pp. 460–464 (2003)
8. Lu, Z., Cohen, K.B., Hunter, L.E.: Finding GeneRIFs via Gene Ontology annotations. In: Pac. Symp. Biocomput., pp. 52–63 (2006)
9. Ding, J., Berleant, D., Nettleton, D., Wurtele, E.: Mining MEDLINE: Abstracts, Sentences, or Phrases? In: Pac. Symp. on Biocomput., vol. 7, pp. 326–337 (2002)
10. Lu, Z.: Text Mining on GeneRIFs. PhD Thesis, Univeristy of Colorado (2007)
11. Blaschke, C., Andrade, M.A., Ouzounis, C., Valencia, A.: Automatic extraction of biological information from scientific text: protein-protein interactions. In: Proc. Int. Conf. Intell. Syst. Mol. Biol., pp. 60–67 (1999)
12. Morgan, A., Lu, Z., Wang, X., Cohen, A., Fluck, J., et al.: Overview of BioCreative II gene normalization. Genome Biol. 9(Suppl. 2), S3 (2008)
13. Apache: Apache UIMA, http://incubator.apache.org/uima/
14. Alias-i. 2008.: LingPipe 3.8.2 (2008), http://alias-i.com/lingpipe/
15. Settles, B.: ABNER: an open source tool for automatically tagging genes, proteins and other entity names in text. Bioinformatics 21(14), 3191–3192 (2005)
16. Hirschman, L., Colosimo, M., Morgan, A., Yeh, A.: Overview of BioCreative task 1B: normalized gene lists. BMC Bioinfo. 6(Suppl. 1), S11 (2005)
17. Alex, B., Grover, C., Haddow, B., Kabadjor, M., Klein, E., Matthews, M., Roebuck, S., Tobin, R., Wang, X.: Assisted Curation: Does Text Mining Really Help? In: Pac. Symp. Biocomput., pp. 556–567 (2008)
18. Leaman, R., Gonzalez, G.: BANNER: An executable survey of advances in biomedical named entity recognition. In: Pac. Symp. Biocomput., vol. 13, pp. 652–663 (2008)

Combining Semantic Relations and DNA Microarray Data for Novel Hypotheses Generation

Dimitar Hristovski[1], Andrej Kastrin[2], Borut Peterlin[2], and Thomas C. Rindflesch[3]

[1] Institute of Biomedical Informatics, Faculty of Medicine, Ljubljana, Slovenia
[2] Institute of Medical Genetics, University Medical Centre, Ljubljana, Slovenia
[3] National Library of Medicine, NIH, Bethesda, MD, USA
dimitar.hristovski@mf.uni-lj.si,
{andrej.kastrin,borut.peterlin}@guest.arnes.si, tcr@nlm.nih.gov

Abstract. Although microarray experiments have great potential to support progress in biomedical research, results are not easy to interpret. Information about the functions and relations of relevant genes needs to be extracted from the vast biomedical literature. A potential solution is to use computerized text analysis methods. Our proposal enhances these methods with semantic relations. We describe an application that integrates such relations with microarray results and discuss its benefits in supporting enhanced access to the relevant literature for interpretation of results and novel hypotheses generation. The application is available at http://sembt.mf.uni-lj.si

Keywords: microarray analysis, literature-based discovery, semantic predications, natural language processing.

1 Introduction

Microarray technology can be used to measure the expression levels of essentially all genes within a genome and can provide insight into gene functions and transcriptional networks [1]. This wealth of information potentially underpins significant advances in biomedical knowledge. However, successful use of microarray data is impossible without comparison to published documents. Due to the large size of the life sciences literature, sophisticated information management techniques are needed to help assimilate online textual resources.

Automatic text mining, commonly based on term co-occurrence, has been used to identify information valuable for interpreting microarray results. In this paper we propose the use of semantic relations (or predications) as a way of extending these techniques. Semantic predications convert textual content into "executable knowledge" amenable to further computation supporting research on genes and relevant diseases. In addition, we suggest that the combination of microarray data and semantic predications can profitably be exploited in the literature-based discovery (LBD) paradigm to further enhance the scientific process.

We describe the use of SemRep [2] for extracting a wide range of semantic predications from MEDLINE citations and discuss a tool for manipulating a database of

C. Blaschke and H. Shatkay (Eds.): ISBM/ECCB 2009, LNBI 6004, pp. 53–61, 2010.

such relations. We then exploit these predications and the results of a microarray experiment from the GEO repository (GSE8397) [3] on Parkinson disease to generate novel hypotheses in the LBD paradigm.

2 Background

A variety of statistical techniques have been used to manipulate text features (usually in MEDLINE citations) to elucidate relevant literature on microarray experiments. Shatkay et al. [4], for example, extract gene function terms from a set of citations identified as related to a kernel document using a document similarity algorithm. Many methods use co-occurring text words [5], often in conjunction with additional information such as MeSH indexing or structured information from related databases such as the Gene Ontology (e.g. [6, 7]). Some systems exploit a thesaurus to identify concepts in text [8] or calculate implicit information by identifying terms related through co-occurrence with shared, intermediate terms [9].

The LBD paradigm was introduced by Swanson [10] for discovering new relations (hypotheses) between concepts by analyzing the research literature. Swanson's method and most of those that followed, including our BITOLA system [11], are co-occurrence based. We expanded the LBD paradigm by using semantic relations and discovery patterns [12], and we applied the expanded methodology to investigate drug mechanisms [13]. In this paper we further expand LBD by combining microarray data with semantic relations extracted from the literature and by defining new discovery patterns.

The SemRep program extracts semantic predications from MEDLINE citations in several domains, including clinical medicine [2], molecular genetics [14], and pharmacogenomics [15]. The system is symbolic and rule based, relying on structured domain knowledge in the Unified Medical Language System® (UMLS),® extended for molecular genetics and pharmacogenomics. SemRep uses underspecified syntactic analysis, in which only simple noun phrases are identified. MetaMap is used [16] to identify Metathesaurus concepts and is augmented by ABGene [17] to identify gene names. Text tokens marked as potential gene names by either MetaMap or ABGene are searched in a precomputed Berkeley DB table compiled from Entrez Gene official symbols, names, aliases, and identifiers. A successful match is given the Entrez Gene identifier. The gene table is updated periodically and is currently limited to human genes.

SemRep predications have Metathesaurus concepts as arguments and Semantic Network relations as predicates. The relations currently addressed are:

Genetic Etiology: ASSOCIATED_WITH, PREDISPOSES, CAUSES

Substance Relations: INTERACTS_WITH, INHIBITS, STIMULATES

Pharmacological Effects: AFFECTS, DISRUPTS, AUGMENTS

Clinical Actions: ADMINISTERED_TO, MANIFESTATION_OF, TREATS

Organism Characteristics: LOCATION_OF, PART_OF, PROCESS_OF

Co-existence: CO-EXISTS_WITH

As an example, SemRep extracts the predication "MDB1 CAUSES Autistic Disorder" from the text ... *the loss of Mbd1 could lead to autism-like behavioral phenotypes* ... In this interpretation, Mbd1 has semantic type 'Gene or Genome' and *autism* maps to the concept "Autistic Disorder" (with semantic type 'Disease or Syndrome'). *Lead to* is an indicator for the semantic relation CAUSES. Similarly the predication "MBD1 INTERACTS_WITH HTR2C" is extracted from ... *Mbd1 can directly regulate the expression of Htr2c, one of the serotonin receptors,* ... on the basis of the identification of the two genes in this text and the verb *regulate* indicating the relation INTERACTS_WITH.

3 Methods

We processed microarray data from the GEO data set and integrated it with SemRep predications in a MySQL database. To accommodate literature-based discovery we formulated discovery patterns [12] that refer to the interaction of drugs and genes. Finally, we devised tools for searching the database using the discovery patterns in order to explore the microarray data and associated research literature for a specific disease and suggest hypotheses about potential drug therapies for that disease.

3.1 Preparing the Microarray Experiments and Results

Currently, we have preprocessed and integrated only a few microarray datasets. In the future we will consider allowing the user to request any GEO dataset be processed with default parameters. Another option is to allow the user to upload a list of differentially expressed genes directly into the system. Below we describe the processing of a GEO dataset that is used throughout the paper to illustrate our methodology and the tools.

A total of 47 Affymetrix HG-U133A CEL files for 29 Parkinson disease patients and 18 controls were retrieved from the GEO repository (GSE8397) [3]. All computations were carried out in the R software environment for statistical computing using additional Bioconductor packages [18, 19]. The normalization of the raw data was performed using the MAS5 algorithm as implemented in the `affy` package. Hybridization probes were mapped to Entrez Gene IDs by annotation data in the `hgu133a.db` package. Analysis of differentially expressed genes (DEG) was performed using Welch's t-test from the `multtest` package. The Benjamini and Hochberg method was selected to adjust p-values for multiple testing [20]. As a confidence threshold we used an adjusted value of $p \leq 0.01$. A total of 567 DEGs were used for further processing.

3.2 Integrated Database with Semantic Relations and Microarray Results

We built an integrated MySQL database to store the semantic relations extracted by SemRep and the microarray results we processed. The data is spread across several tables holding information on the arguments and relations from the predications. For each argument we store concepts and synonyms as well as semantic types. Arguments are UMLS concepts, but when an argument is a gene, in addition to the UMLS CUI (Concept Unique Identifier) we also store the Entrez Gene ID, which serves as a link to the microarray results. In addition, a link is maintained to the sentence in the MEDLINE citation from which the predication was generated.

We have developed two tools for searching the integrated database: one for searching direct relations between concepts and one for indirect relations. In both cases the arguments of the relations can be limited to genes from the microarray. To allow fast and flexible searching of the integrated database we use Lucene and have built separate indexes, one for fast text searching with Lucene and another for accessing the data stored in MySQL when needed. The tools for searching are Web based and were built with the Ruby on Rails application development framework. The tools provide a flexible way to answer questions about what is already known from the literature: genes associated with a disease; relations between a disease and other concepts; relations between the genes from the microarray and themselves or with other concepts. The tools can also generate novel hypotheses: implicit links between a disease and up- or downregulated genes; concepts that might be used to affect these genes; and potential new treatments.

3.3   Discovery Patterns for Novel Hypotheses Generation

For novel hypotheses generation, the tools exploit *discovery patterns,* which are query combinations whose results represent a novel hypothesis – something not specified in the literature or in the microarray results alone. We have designed several new discovery patterns, only two of which are described here. The two discovery patterns, which can be used to discover new therapeutic approaches for some disease, work by regulating the up- or downregulated genes related to that disease (Figure 1).

For example, if we want to investigate regulating genes that are upregulated in the microarray, we search for concepts (genes, drugs, etc.) that are reported in the literature as inhibiting the upregulated genes. We call this discovery pattern "inhibit the upregulated." Similarly, we can investigate downregulated genes with the "stimulate the downregulated" pattern, in which case we search for biomedical concepts that are known to stimulate the downregulated genes.

These discovery patterns combine information from the microarray data about up- or downregulated genes in patients having a certain disease with information from the literature about biomedical concept that can be used to regulate those genes. The discovery patterns can be more complex and involve the combination of more searches through several common intermediate concepts. Also, relations in addition to "INHIBITS" and "STIMULATES," could be used. The novel hypotheses produced by discovery patterns need to be evaluated by a human expert, first by reading the literature and then by laboratory experiments.

Our tools allow complex queries implementing discovery patterns to be specified easily. As output, semantic relations or novel hypotheses are presented first. Then, on request, the highlighted sentences and MEDLINE citations from which the semantic relations are extracted are shown. Some examples are given in the next section.

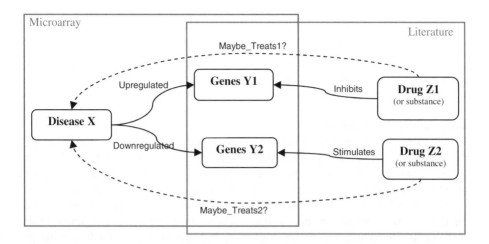

Fig. 1. The two discovery patterns "Inhibit the Upregulated" and "Stimulate the Downregulated" that can be used to find novel therapeutic agents. The patterns combine information from the microarray about which genes are up- or downregulated with information from the literature about which agents could be used to inhibit or stimulate these genes.

4   Results

4.1   Numbers Describing Size of Processing

We used SemRep to process 43,369,616 sentences from 6,699,763 MEDLINE citations published between 1999 and the end of March 2009. 21,089,124 semantic predication instances were extracted, representing 7,051,240 distinct predication types. There are 1,334,014 distinct UMLS concepts appearing as arguments of the semantic predications.

4.2   Evaluation

Evaluating medical aspects of our hypotheses is beyond the scope of this work, and in this paper we do not address the reliability of microarray results. Our focus is on estimating SemRep accuracy, and for this we rely on the work of Masseroli et al [14]. They established a baseline by calculating precision on 2,042 relations extracted with SemGen (now integrated into SemRep): 41.95% for 'genetic' relations (INHIBITS, STIMULATES) and 74.2% for 'etiologic' relations (CAUSES, ASSOCIATED_WITH, PREDISPOSES). They then propose a postprocessing strategy to improve results using the distance (measured in phrases) of the argument (subject and object) from the indicator of the semantic relation. For example, if INHIBITS and STIMULATES relations are filtered for arguments at distance 1 from the indicator, precision increases to 70.75%; however, recall drops to 43.6%. At argument distance 2 (or less) from the indicator, precision is 55.88% and recall 66.28%. In exploiting this method, we first show the user relations more likely to be correct by ranking results in order of increasing argument-predicate distance.

4.3   Generating Novel Hypotheses for Potential Therapeutic Agents

We illustrate the capabilities of our methodology on a microarray for Parkinson disease (PD) (GEO GSE8397) [21] and investigate therapies which might inhibit the expression of upregulated genes or stimulate the expression of downregulated genes associated with this disorder.

4.3.1   Inhibit the Upregulated

Figure 2 shows how the "inhibit the upregulated" pattern is implemented with our tool for searching direct semantic relations from the literature. In the *Query* field we can enter a simple or more complex Boolean query expression. The query terms, by using an appropriate short field name, can refer to the name, semantic type and concept identifier (UMLS CUI or Entrez Gene ID) of the subject and/or object of the semantic relation as well as the name of the relation. In Figure 2 we entered "relation:INHIBITS" which means we want to search for all the biomedical concepts where one of them "INHIBITS" the other. If we select the *Search* button without providing additional constraints we will get the first 20 of about 300,000 "INHIBITS" relations.

To completely implement the "inhibit the upregulated" discovery pattern, we provide an additional constraint in the "Microarray Filter" group of fields. The first field *Experiment* allows us to select the microarray experiment (the default value is *none*). In our case we select a PD experiment denoted here as *Parkinson2* (corresponds to GEO GSE8397). The next field, *Limit arguments,* allows us to select which argument of the semantic relation we want to limit. We have selected *object,* which means that the object of the "INHIBITS" relation must be one of the genes on the selected microarray. The other possibilities for this field are: *any,* meaning we are interested in relations where at least one of the arguments is a gene from the microarray; *subject,* meaning the subject of the relations has to be one of the microarray genes; and *both,* meaning only direct relations between the genes on the microarray are to be retrieved.

The next two fields allow us to specify the number of microarray genes to be used for filtering. Here *top N* refers to the most differentially expressed genes. We can select only the *upregulated* or the *downregulated* or *any,* meaning the top N up- or downregulated. Because of performance and implementation issues the top N currently can not be more than 400 genes. The final field in this group allows us to select genes based on the p value. The upper part of Figure 2 shows the options specified for the following example.

As a result of the query we get a list of semantic relations ordered by ascending frequency. For each relation, the subject, the relation itself, the object, and frequency of occurrence are shown. Frequency of occurrence indicates the number of sentences from which the semantic relation was extracted. The frequency number is actually a hyperlink which can be selected to show the list of sentences from which the relations were extracted (subject, relation, and object are highlighted). Additionally, the PubMed ID (PMID) is provided for each sentence; this can be selected to show the PubMed citation in which the sentence appears. Examples of this are shown below.

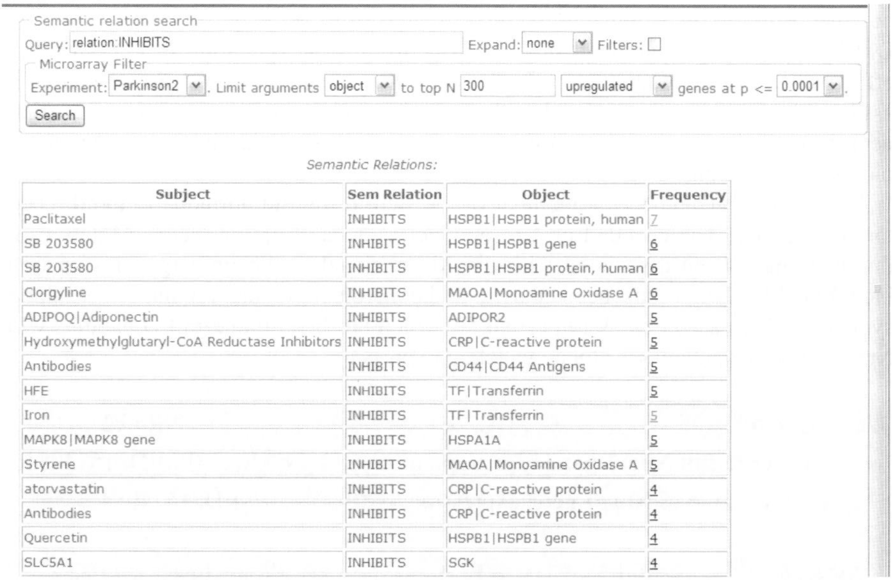

Fig. 2. Finding agents that inhibit some of the genes that are upregulated on a particular Parkinson disease microarray

The HSP27 (HSPB1) gene, which is over-expressed in the experimental results, has already been implicated in the pathogenesis of PD [22]. We identified paclitaxel and quercetin as substances that inhibit the expression of this gene. Paclitaxel has been identified and used as an antineoplastic agent due to its unique activity as a microtubule-stabilizing agent. Interestingly, microtubules appear to be critical for the survival and death of nigral DA neurons, which are selectively affected in PD. Quercetin is a multipotent bioflavonoid with great potential for the prevention and treatment of disease. There is evidence of various *in vivo* and *in vitro* effects of quercetin, including anti-inflamatory, antioxidative, and potentially antineurodegenerative effects relevant to PD.

Paclitaxel	INHIBITS	HSPB1|HSPB1 protein, human

Paclitaxel inhibits expression of **heat shock protein 27** (PMID: 15304155)

Paclitaxel (Pacl) was reported to suppress **HSP27** (PMID: 19080259)

Quercetin	INHIBITS	HSPB1|HSPB1 gene

Quercetin ..., inhibited the expression of both HSP70 and **HSP27** (PMID: 12926076)

4.3.2 Stimulate the Downregulated
Our approach also provides interesting results when we search for substances that stimulate downregulated genes in the transcriptomic experiment. For example, it turns

out that *Pramipexol* stimulates expression of *NR4A2*. Pramipexol is a non-ergotic D2/D3 dopaminergic agonist that can be used to treat the symptoms of PD safely and effectively, both as monotherapy in the early stages and in the advanced phases in association with levodopa. Furthermore, in laboratory studies pramipexole exerts neuroprotective effects and its use has been related to a delay in the appearance of motor complications.

NR4A2 (Nurr1) encodes a member of the steroid-thyroid hormone-retinoid receptor superfamily. The encoded protein may act as a transcription factor. Mutations in this gene have been associated with disorders related to dopaminergic dysfunction, including PD. Nurr1 has been shown to be involved in the regulation of alpha-synuclein. Decreased expression of Nurr1, which has been found in PD patients with Nurr1 mutations, increases alpha-synuclein expression.

pramipexol	STIMULATES	NR4A2
... the increase of **Nurr1 gene** expression induced by PRX, ... (PMID: 15740846)		
... the induction of **Nurr1 gene** expression by PRX ... (PMID: 15740846)		

NR4A2	ASSOCIATED_WITH	Parkinson Disease
... lower levels of **NURR1** gene expression were associated with significantly increased risk for PD (PMID: 18684475)		

We also found that *leptin* stimulates *CDC42*. This gene, which is downregulated in the transcriptomic experiment, codes for a protein which is a small GTPase. Recent data indicate that components of small GTPase signal transduction pathways may be directly targeted by alpha-synuclein oligomers, which potentially leads to signaling deficits and neurodegeneration in PD. Leptin on the other hand is a hormone secreted from white adipocyts. There is evidence that leptin prevents the degeneration of dopaminergic neurons by 6-OHDA and may be useful in treating PD.

5 Conclusion

In this paper we presented an application that integrates the results of microarray experiments with a large database of semantic predications representing the content of nearly 5 million MEDLINE citations. We discuss the value of this system with examples from microarray data on Parkinson disease, illustrating the way semantic relations elucidate the relationship between current knowledge and information gleaned from the experiment and help generate novel hypotheses.

References

1. Cordero, F., Botta, M., Calogero, R.A.: Microarray data analysis and mining approaches. Brief Funct. Genomic Proteomic 6, 265–281 (2007)
2. Rindflesch, T.C., Fiszman, M.: The interaction of domain knowledge and linguistic structure in natural language processing: interpreting hypernymic propositions in biomedical text. J. Biomed. Inform. 36, 462–477 (2003)

3. Barrett, T., Troup, D.B., Wilhite, S.E., Ledoux, P., Rudnev, D., Evangelista, C., Kim, I.F., Soboleva, A., Tomashevsky, M., Edgar, R.: NCBI GEO: Mining tens of millions of expression profiles - database and tools update. Nucleic Acids Res. 35 (Database issue), D760–D765 (2007)
4. Shatkay, H., Edwards, S., Wilbur, W.J., Boguski, M.: Genes, themes and microarrays: using information retrieval for large-scale gene analysis. In: Proc. Int. Conf. Intell. Syst. Mol. Biol., pp. 317–328 (2000)
5. Blaschke, C., Oliveros, J.C., Valencia, A.: Mining functional information associated with expression arrays. Funct. Integr. Genomics 1, 256–268 (2001)
6. Yang, J., Cohen, A.M., Hersh, W.: Automatic summarization of mouse gene information by clustering and sentence extraction from MEDLINE abstracts. In: AMIA Annu. Symp. Proc., pp. 831–835 (2007)
7. Leach, S.M., Tipney, H., Feng, W., Baumgartner, W.A., Kasliwal, P., Schuyler, R.P., Williams, T., Spritz, R.A., Hunter, L.: Biomedical discovery acceleration, with applications to craniofacial development. PLoS Comput. Biol. 3, e1000215 (2009)
8. Jelier, R., 't Hoen, P.A., Sterrenburg, E., den Dunnen, J.T., van Ommen, G.J., Kors, J.A., Mons, B.: Literature-aided meta-analysis of microarray data: a compendium study on muscle development and disease. BMC Bioinformatics 9, 291 (2008)
9. Burkart, M.F., Wren, J.D., Herschkowitz, J.I., Perou, C.M., Garner, H.R.: Clustering microarray-derived gene lists through implicit literature relationships. Bioinformatics 23, 1995–2003 (2007)
10. Swanson, D.R.: Fish oil, Raynaud's syndrome, and undiscovered public knowledge. Perspect Biol. Med. 30, 7–18 (1986)
11. Hristovski, D., Peterlin, B., Mitchell, J.A., Humphrey, S.M.: Using literature-based discovery to identify disease candidate genes. Int. J. Med. Inform. 74, 289–298 (2005)
12. Hristovski, D., Friedman, C., Rindflesch, T.C., Peterlin, B.: Exploiting semantic relations for literature-based discovery. In: AMIA Annu. Symp. Proc., pp. 349–353 (2006)
13. Ahlers, C.B., Hristovski, D., Kilicoglu, H., Rindflesch, T.C.: Using the literature-based discovery paradigm to investigate drug mechanisms. In: AMIA Annu. Symp. Proc., pp. 6–10 (2007)
14. Masseroli, M., Kilicoglu, H., Lang, F.M., Rindflesch, T.C.: Argument-predicate distance as a filter for enhancing precision in extracting predications on the genetic etiology of disease. BMC Bioinformatics 7, 291 (2006)
15. Ahlers, C.B., Fiszman, M., Demner-Fushman, D., Lang, F.M., Rindflesch, T.C.: Extracting semantic predications from Medline citations for pharmacogenomics. In: Pac. Symp. Biocomput., pp. 209–220 (2007)
16. Aronson, A.R.: Effective mapping of biomedical text to the UMLS Metathesaurus: The MetaMap program. In: Proc. AMIA Symp., pp. 17–21 (2001)
17. Tanabe, L., Wilbur, W.J.: Tagging gene and protein names in biomedical text. Bioinformatics 18, 1124–1132 (2002)
18. R Development Core Team.: R: A language and environment for statistical computing. R Foundation for Statistical Computing, Vienna, Austria (2008)
19. Gentleman, R.C., et al.: Bioconductor: Open software development for computational biology and bioinformatics. Genome Biol. 5, R80 (2004)
20. Benjamini, Y., Hochberg, Y.: Controlling the false discovery rate: A practical and powerful approach to multiple testing. J. Roy Stat. Soc. B 57, 289–300 (1995)
21. Moran, L.B., Duke, D.C., Deprez, M., Dexter, D.T., Pearce, R.K., Graeber, M.B.: Whole genome expression profiling of the medial and lateral substantia nigra in Parkinson's disease. Neurogenetics 7, 1–11 (2006)
22. White, L.R., Toft, M., Kvam, S.N., Farrer, M.J., Aasly, J.O.: MAPK-pathway activity, Lrrk2 G2019S, and Parkinson's disease. J. Neurosci. Res. 85, 1288–1294 (2007)

Learning from Positive and Unlabeled Documents for Retrieval of Bacterial Protein-Protein Interaction Literature

Hongfang Liu[1], Manabu Torii[2], Guixian Xu[1,4], Zhangzhi Hu[3],
and Johannes Goll[5]

[1] Department of Biostatistics, Bioinformatics, and Biomathematics
[2] Imaging Science and Information Systems Center
[3] Department of Oncology, Georgetown University Medical Center, Washington DC
[4] School of Computer Science and Technology, Beijing Institute of Technology
[5] The J. Craig Venter Institute, Rockville, Maryland
{hl224,mt352,gx6,zh9}@georgetown.edu, jgoll@jcvi.org

Abstract. With the advance of high-throughput genomics and proteomics technologies, it becomes critical to mine and curate protein-
protein interaction (PPI) networks from biological research literature.
Several PPI knowledge bases have been curated by domain experts but
they are far from comprehensive. Observing that PPI-relevant documents
can be obtained from PPI knowledge bases recording literature evidences
and also that a large number of unlabeled documents (mostly negative)
are freely available, we investigated *learning from positive and unlabeled
data (LPU)* and developed an automated system for the retrieval of
PPI-relevant articles aiming at assisting the curation of a bacterial PPI
knowledge base, MPIDB. Two different approaches of obtaining unlabeled documents were used: one based on PubMed MeSH term search
and the other based on an existing knowledge base, UniProtKB. We
found unlabeled documents obtained from UniProtKB tend to yield better document classifiers for PPI curation purposes. Our study shows that
LPU is a possible scenario for the development of an automated system
to retrieve PPI-relevant articles, where there is no requirement for extra
annotation effort. Selection of machine learning algorithms and that of
unlabeled documents would be critical in constructing an effective LPU-
based system.

Keywords: document retrieval, learning from positive and unlabeled,
protein-protein interaction.

1   Introduction

Protein-protein interaction (PPI) knowledge is essential in understanding the
fundamental processes governing cell biology, and it promises to reveal functional
clues about genes in the context of their interacting partners. Recently, large
scale studies of PPI networks have become possible due to advances in experimental high-throughput genomics and proteomics technologies. Experimental

C. Blaschke and H. Shatkay (Eds.): ISBM/ECCB 2009, LNBI 6004, pp. 62–70, 2010.
© Springer-Verlag Berlin Heidelberg 2010

PPI data has been generated for different organisms, and it has been reported in peer-reviewed literature or stored in PPI knowledge bases. The amount of data currently available in PPI knowledge bases is far from comprehensive [1], and there is a large body of PPI data buried in literature. Automated document classification has a potential to accelerate the literature-based curation of a PPI database by domain experts [2,3,4,5]. A machine learning system can classify documents as PPI-related or not, i.e., positives or negatives, respectively. However, development of machine learning classifiers requires a large amount of annotated data. Obtaining such data can be difficult and time-consuming. Sometimes positive and unlabeled examples are readily available in a curation project but negative examples cannot be obtained without paying an additional cost. With more and more such situations in real-life applications, learning from positive and unlabeled data (LPU) has become an important research topic. In this paper, we report our exploration of using two different approaches to obtaining unlabeled documents when building a document retrieval system to assist the curation efforts of the Microbial Protein Interaction Database (MPIDB) [6]. In the following, we first describe the background of classification algorithms. The experimental methods are introduced next. We then present the results and discussion, and conclude our work.

2 Background and Related Work

Recently, Elkan and Noto showed that if positive documents were randomly sampled, the predicted probabilities of being positive by a classifier trained on positive and unlabeled differs from the true conditional probabilities of being positive only by a constant factor [7]. Their study provides a justification of LPU. Some LPU systems were constructed by considering unlabeled documents as negatives. For example, Noto et al. considered the retrieval of papers related to trans-membrane transport-related proteins as an LPU task and used SVMs to rank them according to their relevancy to the topic [8]. Some LPU systems adopt a two-step strategy where the first step is to obtain a reliable negative data set (RN) based on keywords or available classifiers and the second step is to refine or augment RN using various learning approaches such as clustering or boosting. One popular approach to obtain reliable negatives is to define a classification task which considers unlabeled documents as negatives, builds a binary classifier, and treats those classified as negatives to be RN. For example, Li and Liu [9] proposed a method that first builds a Rocchio classifier based on positives, and documents classified as negatives are treated as RN. Similar technique can be used to augment training data for PPI document classification. For example, Tsai et al. [10] trained an SVM classifier with 3,536 positives and 1,959 negatives manually annotated for the BioCreative 2 workshop. Then the training set was augmented with likely positives and negatives selected among 50,000 unlabeled documents using the initial SVM. Our prior work [11] also indicates that RN can be obtained by considering unlabeled documents as negatives.

Related studies focus on various LPU methods and machine learning algorithms. In this paper, we report our findings of using different unlabeled data sets to train LPU classifiers to assist the curation of MPIDB, a knowledge base that has been actively curated for microbial protein interactions [6]. A set of positive documents was obtained from MPIDB and two sets of unlabeled documents were tested: one obtained from PubMed using the MeSH term *bacterial protein*, and the other obtained based on UniProtKB cross-references as detailed below. We compared the performance of classifiers developed using these two unlabeled document collections.

3   Experimental Methods

There are different ways to obtain unlabeled documents for the training of LPU systems, and the selection of unlabeled document sets will affect performance of resulting systems. In this study, we compared two approaches of obtaining unlabeled documents for training machine learning classifiers to assist manual curation of bacterial PPI for MPIDB. The following describes the experiment in detail.

3.1   Data Acquisition

We used MPI-LIT, a curated subset of MPIDB [12], and obtained 814 positive documents (abstracts) from 31 journals, such as *Journal of bacteriology*, *Journal of molecular biology*, and *Molecular microbiology*. Unlabeled documents were gathered using two different approaches. One is to retrieve MEDLINE abstracts published in these 31 journals (accessed on Feb 1, 2009) within the last five years and indexed with the MeSH term *bacterial proteins*. After removing positive articles, we obtained 29,308 unlabeled documents (denoted as UL_5Years). The other is to retrieve cross-references from UniProtKB records for organisms included in MPIDB (a total of 93 organisms). After removing positives, we obtained 27,140 unlabeled documents (denoted as UL_UniProtKB). There are 5,078 documents in both unlabeled document collections (denoted as UL_Overlap). Figure 1 provides the overview of the unlabeled data sets.

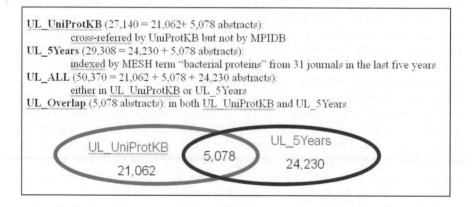

Fig. 1. The overview of the unlabeled data sets

3.2   Data Preprocessing

We normalized the text by changing nouns in plural forms into singular forms, verbs in past tense into present tense, and replacing nouns and adjectives by their corresponding verbs based on the SPECIALIST lexicon, a component in the Unified Medical Language System (UMLS) [13]. We also replaced punctuation marks with spaces and changed uppercase letters to lowercase ones. For example, the sentence *This interaction results in the activation of caspase-1, as seen in its proteolytic maturation and the processing of its substrate interleukin-1beta.* is normalized to *this interact result in the activate of caspase 1 as see in its proteolytic mature and the process of its substrate interleukin 1 beta.*

3.3   Machine Learning Algorithms

A growing number of machine learning algorithms have been applied to document classification. Examples include K nearest neighbor, Bayesian approaches, decision trees, symbolic rule learning, support vector machines (SVM) and neural networks [14,15,16,17]. We limited the selection of machine learning algorithms to five algorithms: three implemented in the Rainbow software, i.e., Rocchio, Naive Bayes (NB), and a probabilistic indexing classifier (PRIND) [18] and two other widely used classification algorithms: SVM and Logistic Regression (LR). The following summarizes those machine learning algorithms.

Rocchio is a similarity-based classification method. Documents are represented as feature vectors where features are words in the documents and feature values are calculated based on *TF-IDF (term frequency-inverse document frequency)* (see, e.g., [19]). Each class is then represented as a feature vector computed as the vector average of its members. The predicted class of a document is one yielding the highest similarity measured as the cosine of the document feature vector and class feature vector. Naive Bayes (NB) is a probabilistic based approach which applies Bayes' theorem to classify documents. The predicted class C' of a document d' is one with the highest conditional probability computed by assuming that features occur independently, i.e., $P(C'|d') \propto P(C') \prod_w P(w|C')^{TF(w,d')}$ where $P(\cdot)$ is the probability and $TF(w, d')$ is the frequency of w in d'. PRIND [19] is also a probabilistic based approach but based on *retrieval with probabilistic indexing* to classify documents. The predicted class C' of a document d' is one with the highest value of the following formula: $\sum_w \frac{P(w|C')P(w|d')P(C')}{\sum_{C_j} P(w|C_j)P(C_j)}$. SVM [20] is a kernel-based method and seeks a hyper-plane in the feature space that maximizes the margin between the two sets of documents. The so-called *kernel trick* is used to transform features into higher space to seek separable hyper-plane in the feature space. Past studies suggest that a linear classifier is usually sufficient for text data, e.g., [21]. The derivation of a hyper-plane is a numerical optimization process that can be computationally expensive, but efficient implementations of SVM have been publicly available, e.g., LibSVM. An output from an SVM classifier reflects the distance of an instance from a derived hyper-plane, but calibration of probabilities for predicted classes has also been studied for SVM [22,23]. Logistic Regression (LR) is used for prediction of binary-event probabilities by fitting a generalized linear model to provided data points. Like many forms of regression analysis, it makes use of several predictor variables that may be either numerical or categorical. LR can be expensive in high-dimension learning but recently with the availability of fast implementations, it has gained popularity for high-dimensional classification such as text processing where linear boundaries are usually adequate to separate the classes [24].

3.4   Experiment Design and Evaluation

We used the following approach to obtain reliable negatives. Given a training set (TR) consisting of positive and unlabeled documents, we assumed the latter as negative documents and constructed five classifier models using the default settings in Rainbow, one for each of the machine learning algorithms (i.e., Rocchio, Naive Bayes, PRIND, SVM, and LR). The classifiers were applied to unlabeled documents in the training set. We ordered the documents classified as negatives into rank lists based on scores by the classifiers in which articles ranked higher were more likely to be negatives. We then combined the five rank lists to obtain RN (reliable negative data set) using the following approach:

- For each document, the highest rank assigned to it among the five rank lists was used as the new score of the document;
- A new rank list was generated based on the new score; and
- Those top ranked documents are included as RN where the ratio of the number of documents in RN to the number of positives in TR is controlled to be 10:1, which was selected after trial-and-error.

After obtaining RN, a classifier was constructed based on positives in TR and predicted negatives in RN. Three unlabeled collections were compared: UL_UniProtKB (unlabeled extracted from UniProtKB), UL_5Years (unlabeled extracted from PubMed), and UL_ALL (those in either of the two collections). The performance was measured using Area under ROC curve (AUC) based on positives and UL_Overlap (i.e, those in both UL_UniProtKB and UL_5Years). Given a ranked list where predicted positives were ranked highly, AUC is interpreted as the probability that the rank of a positive document d_1 is greater (i.e., more likely to be positive) than that of an unlabeled document d_0, where d_1 and d_0 are documents randomly selected from positive and unlabeled document sets, respectively. We assume the higher the AUC value, the better the classifier is (see the details in [25]). The performance was measured using 5-fold cross validation.

4   Results and Discussions

Table 1 shows the AUC results. The fifteen ROCs are shown in Figure 2. Each plot in the left panel of Figure 2 shows the performance of five machine learning algorithms for a given data set and each plot in the right panel of Figure 2 compares the performance of different data sets for a given machine learning

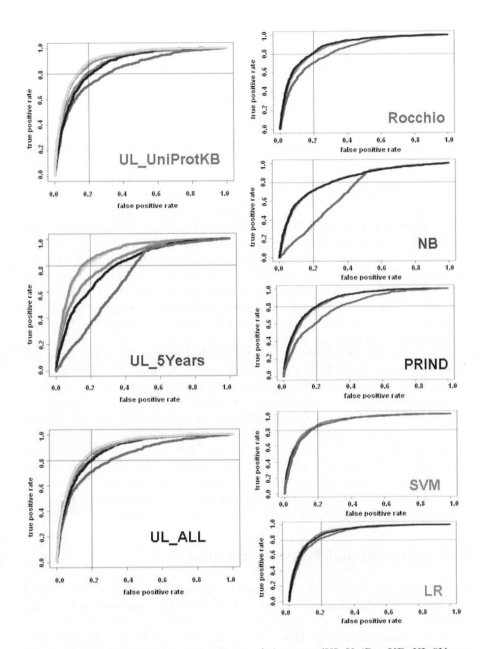

Fig. 2. Fifteen ROCs arranged according to i) data sets (UL_UniProtKB, UL_5Years, and UL_ALL) shown in the left panel, and ii) machine learning algorithms (Rocchio, NB, PRIND, SVM, and LR). The legend for plots in the left panel is (Rocchio - red, NB - blue, PRIND - black, SVM - purple, and LR - gray) and the legend for plots in the right panel is (UL_UniProtKB -red, UL_5Years - blue and UL_ALL - black).

Table 1. AUC results (obtained based on 814 positive and 5,078 unlabeled documents)

Data set	Rocchio	NB	PRIND	SVM	LR
UL_UniProtKB	87.9	82.8	87.3	90.9	90.9
UL_5Years	84.0	71.3	80.5	90.4	90.4
UL_All	88.9	82.9	88.1	90.0	90.5

algorithms. Note that our definition of AUC is based on positives and unlabeled and the real performance measure can be higher since some of the high-ranked unlabeled were found to be true positives. From Table 1 and Figure 2, we can see that SVM and LR were comparable to each other and were superior to Rocchio, NB, and PRIND. Among three algorithms from Rainbow, Rocchio on UL_All achieved the best AUC (AUC=.889) and NB on UL_5Years performed the worst (AUC=.713). Note that NB tended to assign scores of 1.0 for either positives or unlabeled documents when using UL_5Years as the unlabeled data set. Such phenomenon is caused by the independence assumption underlying NB, which has been reported previously [26], and it could be affected by the skewed class distribution as well. Additionally, we found that SVM and LR were less sensitive to the selection of the unlabeled data set and achieved comparable performance among the three data sets.

The results also indicated that UL_UniProtKB achieved significantly better performance than unlabeled documents retrieved from PubMed for Rocchio, NB, and PRIND. One possible reason is because documents cross-referred by UniProtKB but not included in MPIDB are mostly negative (i.e., not PPI-relevant), i.e., documents cross-referred by UniProtKB have been mostly reviewed for curation of MPIDB in the past, while documents retrieved for MeSH term *bacterial proteins* but not included in MPIDB still contain more positives. Additionally, UL_ALL and UL_UniProtKB tended to achieve similar performance even UL_ALL has much more unlabeled documents than UL_UniProtKB. It indicates that it is not true that more unlabeled data yield better LPU systems and the appropriate choice of unlabeled documents is important for deriving a good LPU system.

5   Conclusion and Future Direction

In this paper, we report the use of manually curated articles from the MPIDB and unlabeled documents from MEDLINE to build document retrieval systems aiming at accelerating the curation of bacterial protein interactions. We compared and evaluated five machine learning algorithms and two unlabeled document collections for the retrieval of bacterial PPI-relevant documents. Our study indicates that the appropriate selection of unlabeled documents is important in developing LPU systems, and machine learning classifiers trained on document collections compiled from an existing knowledge base tend to perform better than those trained on unlabeled documents retrieved from PubMed based on a MeSH term.

Document retrieval systems based on positives and unlabeled documents achieved performance acceptable by MPIDB curators for practical database curation. It is not clear, however, if LPU systems are competitive with or superior to regular supervised systems trained on positives and a limited number of manually annotated negatives that could be obtained without disturbing the regular curation task. We plan to compare LPU and regular supervised machine learning systems in our future study.

Acknowledgement

We thank anonymous reviewers for their suggestions and very helpful comments. We would also like to acknowledge Drs. Linda Hannick and Shelby Bidwell for their curation effort. This project is supported by DBI-0845523 from the National Science Foundation.

References

1. Morrison, J.L., Breitling, R., Higham, D.J., Gilbert, D.R.: GeneRank: using search engine technology for the analysis of microarray experiments. BMC Bioinformatics 6, 233 (2005)
2. Spasic, I., Ananiadou, S., McNaught, J., Kumar, A.: Text mining and ontologies in biomedicine: making sense of raw text. Brief Bioinform. 6, 239–251 (2005)
3. Leitner, F., Krallinger, M., Rodriguez-Pebagosa, C., et al.: Introducing Meta-Services for Biomedical Information Extraction. Genome Biology (2009) (in press)
4. Krallinger, M., Morgan, A., Smith, L., Leitner, F., Tanabe, L.: Evaluation of text mining systems for biology: overview of the Second BioCreAtIve community challenge. Genome Biology 9(Suppl. 2), S1 (2008)
5. Krallinger, M., Valencia, A., Hirschman, L.: Linking genes to literature: text mining, information extraction, and retrieval applications for biology. Genome Biol. 9(Suppl. 2), S8 (2008)
6. Goll, J., Rajagopala, S.V., Shiau, S.C., Wu, H., Lamb, B.T., Uetz, P.: MPIDB: the microbial protein interaction database. Bioinformatics 24, 1743–1744 (2008)
7. Elkan, C., Noto, K.: Learning classifiers from only positive and unlabeled data. In: Proceedings of the Fourteenth International Conference on Knowledge Discovery and Data Mining, KDD (2008)
8. Noto, K., Saier Jr., M.H., Elkan, C.: Learning to find relevant biological articles without negative training examples. In: Proceedings of the 21st Australasian Joint Conference on Artificial Intelligence, AI (2008)
9. Li, X., Liu, B.: Learning to classify text using positive and unlabeled data. In: Proceedings of Eighteenth International Joint Conference on Artificial Intelligence (2003)
10. Tsai, R.T., Hung, H.C., Dai, H.J., Lin, Y.W., Hsu, W.L.: Exploiting likely-positive and unlabeled data to improve the identification of protein-protein interaction articles. BMC Bioinformatics 9(Suppl. 1), S3 (2008)
11. Xu, G., Niu, Z., Uetz, P., Gao, X., Qin, X., Liu, H.: Semi-Supervised Learning of Text Classification on Bacterial Protein-Protein Interaction Documents. Presented at International Joint Conference on Bioinformatics, Systems Biology and Intselligent Computing, IJCBS 2009 (2009)

12. Rajagopala, S.V., Goll, J., Gowda, N.D., Sunil, K.C., Titz, B., Mukherjee, A., Mary, S.S., Raviswaran, N., Poojari, C.S., Ramachandra, S.: MPI-LIT: A literature-curated dataset of microbial binary protein-protein interactions. Bioinformatics (2008)

13. Bodenreider, O.: The Unified Medical Language System (UMLS): integrating biomedical terminology. Nucleic Acids Res. 32, D267–D270 (2004)

14. Mladenic, D.: Feature subset selection in text learning. In: Nédellec, C., Rouveirol, C. (eds.) ECML 1998. LNCS, vol. 1398, pp. 95–100. Springer, Heidelberg (1998)

15. Lewis, D.D., Ringuette, M.: A comparison of two learning algorithms for text categorization. In: Proceedings of SDAIR 1994, 3rd Annual Symposium on Document Analysis and Information Retrieval, pp. 81–93 (1994)

16. Cohen, W.W., Singer, Y.: Context-Sensitive Learning Methods for Text Categorization. ACM Transactions on Information Systems (TOIS) 17, 141–173 (1999)

17. Wiener, E.D., Pedersen, I.O., Weigend, A.S.: A neural network approach to topic spotting. In: Proceedings of SDAIR 1995, 4th Annual Symposium on Document Analysis and Information Retrieval, pp. 317–332 (1995)

18. McCallum, A.K.: Bow: A toolkit for statistical language modeling, text retrieval, classification and clustering, http://www-2.cs.cmu.edu/~mccallum/bow/

19. Joachims, T.: A Probabilistic Analysis of the Rocchio Algorithm with TFIDF for text Categorization. In: Proceedings of the Fourteenth International Conference on Machine Learning, pp. 143–151 (1997)

20. Vapnik, V.N.: The Nature of Statistical Learning Theory. Springer, Heidelberg (2000)

21. Joachims, T.: Text categorization with Support Vector Machines: Learning with many relevant features. In: Nédellec, C., Rouveirol, C. (eds.) ECML 1998. LNCS, vol. 1398, pp. 137–142. Springer, Heidelberg (1998)

22. Chang, C.-C., Lin, C.-J.: LIBSVM: A library for support vector machines, http://www.csie.ntu.edu.tw/~cjlin/libsvm/

23. Scholkopf, B., Smola, A.J.: Learning with Kernels: Support Vector Machines, Regularization, Optimization, and Beyond. MIT Press, Cambridge (2001)

24. Komarek, P., Moore, A.: Making logistic regression a core data mining tool: A practical investigation of accuracy, speed, and simplicity, pp. 685–688. Carnegie Mellon University, Pittsburgh (2005)

25. Hand, D.J., Mannila, H., Smyth, P.: Principles of Data Mining. MIT Press, Cambridge (2001)

26. Bennett, P.N.: Assessing the calibration of Naive Bayes posterior estimates. Technical Report, CMU-CS-00-155, School of Computer Science. Carnegie-Mellon University, Pittsburgh (2000)

Extracting and Normalizing Gene/Protein Mentions with the Flexible and Trainable Moara Java Library

Mariana L. Neves, José Maria Carazo, and Alberto Pascual-Montano

Biocomputing Unit, Centro Nacional de Biotecnología – CSIC,
C/ Darwin 3, Campus de Cantoblanco, 28049, Madrid, Spain
{mlara,carazo,pascual}@cnb.csic.es

Abstract. Gene/protein recognition and normalization are important prerequisite steps for many biological text mining tasks. Even if great efforts have been dedicated to these problems and effective solutions have been reported, the availability of easily integrated tools to perform these tasks is still deficient. We therefore propose Moara, a Java library that implements gene/protein recognition and normalization steps based on machine learning approaches. The system may be trained with extra documents for the recognition procedure and new organism may be added in the normalization step. The novelty of the methodology used in Moara lies in the design of a system that is not tailored to a specific organism and therefore does not need any organism-dependent tuning in the algorithms and in the dictionaries it uses. Moara can be used either as a standalone application or incorporated in a text mining system and it is available at: http://moara.dacya.ucm.es

Keywords: biomedical text mining, gene/protein recognition and normalization, machine learning.

1 Introduction

Some of the most important steps in the process of analysis of scientific literature are related to the extraction and identification of genes and proteins in the text and their association to the particular entry in their corresponding biological database. This is known as gene/protein recognition and normalization and they are common prerequisite tasks to some more complex text mining systems.

The main difficulties of the gene/protein recognition and normalization problems lies in the high number of existing genes and proteins entities along with the lack of standards in their nomenclature. Some of these entities coincide with common English words, which make their detection in text very complex. Also, nomenclature may appear as long descriptive names or as acronyms, which make identification even more difficult. This situation gets worse since existing biological entities may also have their original name changed. The new discovered biological entities only aggravate this problem since their assigned name may be the same of an existing one.

In the case of the gene normalization task, different species usually need different strategies depending on the complexity of the nomenclature and the degree of ambiguity in their synonyms and even among organisms, because the same mention may

C. Blaschke and H. Shatkay (Eds.): ISBM/ECCB 2009, LNBI 6004, pp. 71–80, 2010.
© Springer-Verlag Berlin Heidelberg 2010

refer to distinct entities of the same or distinct species. Due to its importance, the gene/protein recognition and normalization tasks have received a lot of attention from the scientific community. One clear example is the BioCreative challenges [1-3], a community-wide effort for evaluating text mining systems in the biological domain. Despite the great efforts of the scientific community in the improvement of these tasks, the availability of reliable systems and dictionaries of synonyms that can be easily integrated in more general text mining systems is still deficient.

In this paper we introduce Moara, which comes as a freely available Java library alternative. Moara has been running for more than one year and improvements have been made regarding new functionalities and a more stable version. The gene/protein recognition and normalization tasks are carried out by a Case-Based Reasoning [4] approach (CBR-Tagger) and a mix of few organism-dependent knowledge and machine learning methodologies (ML-Normalization), respectively. The latter is available for four organisms: *Saccharomyces cerevisiae* (yeast), *Mus musculus* (mouse), *Drosophila melanogaster* (fly) and *Homo sapiens* (human). However, the system may be trained with new organisms.

Moara can use different sources of information to train the system for a given organism and uses a methodological strategy that proves to be effective in several scenarios and that can be extended to increase its performance. The system makes use of some MySQL databases and two external libraries: Weka machine learning tool [5] and SecondString[1] [6] for the string distances. Detailed documentation as well as code examples is presented at the documentation page[2]. Additional results of our experiments are presented at the supplementary material page[3].

2 Related Work

The gene/protein mention recognition problem has been the focus of research of the bioinformatics community during the last few years, therefore various methodologies have already been proposed for this problem. This type of biological entities is usually identified in a text by means of part-of-speech tags and orthographic features. Several algorithms that make use of the above features have been presented. For example, manual rules [7], logistic regression [8], Conditional Random Fields [9] and machine learning [10]. Among them we draw attention to the freely available Abner [11] and Banner [12] taggers that we use in this work. Both of them make use of Conditional Random Fields and orthographic and contextual features. Abner and Banner report an F-Measure of 69.9 (BioCreative 1 dataset) and 84.92 (BioCreative 2 dataset), respectively.

Although there are many freely available taggers, a mix of them is desirable in order to increase the recall and be able to extract most of the mentions from a text. We think that the more taggers we use the better coverage we would have as this situation is equivalent to the combination of classification methods that is commonly used in machine learning. We have used only two external taggers since those are well-known

---

[1] http://secondstring.sourceforge.net/
[2] http://moara.dacya.ucm.es/documentation.html
[3] http://moara.dacya.ucm.es/suppl_material_lncs.html

and accepted by the community and also easily integrated to our system, although more taggers can also be included.

Many solutions have also been proposed to the gene normalization problem and most of them share a similar sequence of steps that consist in first extracting the gene/protein mentions from the text followed by a matching procedure against a pre-processed dictionary of synonyms for each organism. An optional last step includes filtering the results and/or disambiguating the candidates' identifiers, in case that more than one is found for the same mention.

The first step of extracting the mentions from the text can be carried out by the same system responsible for the normalization task [13], or accomplished by one or more of the available taggers. The matching procedure use either an exact or a rule-based approach operating on an curated dictionary of synonyms [13] or an approximated matching [14] such as a string similarity approach based on Jaro-Winkler [6]. The dictionary of synonyms is usually constructed automatically by joining information from many online biological databases [15] and the synonyms may be pruned automatically [14] or manually by experts [13].

The last and sometimes optional steps of the gene normalization problem are the disambiguation and filtering tasks. Different approaches such as context information [16], machine learning based filters [13] and similarity measure between the text and disambiguation vectors for each gene [15, 17]. Farkas [18] has proposed to use the influence of the co-authors in the nomenclature of the entities.

3 Methods

3.1 CBR-Tagger

The gene/protein recognition is carried out by the CBR-Tagger [19], a tagger based on Case-Based Reasoning (CBR) foundations [4] that in its initial version [20] participated in the BioCreative 2 Gene Mention task [1]. CBR-Tagger comes with five different models according to the datasets that have been used in the training step: the BioCreative 2 GM [1] alone (CbrBC2) and the latter combined with the BioCreative task 1B [2] corpora for the yeast (CbrBC2y), mouse (CbrBC2m), fly (CbrBC2f) and the three of them (CbrBC2ymf), in order to be able to better extract mentions from the referred organisms. CBR-Tagger may be trained with extra corpora and the only requirement is that documents should be provided in the format used in the BioCreative 2 GM task [1]. In addition, it is possible to use the cases that have been already learned from the five models previously discussed.

CBR is a machine learning method that consists of first learning cases from the training documents and saving them in a base of cases. During the testing step, the case most similar to a given situation is retrieved from the base (case-solution), from which the final solution will be provided. One of the advantages of CBR is the possibility of getting an explanation of why a certain category is assigned to a given token by checking the features that compose the case-solution. Also, the base of cases may be used as source of knowledge to learn extra information about the training dataset, such as the number of tokens (or cases) that share a certain value of a feature.

In a first step, several cases of the two classes (gene/protein mention or not) are stored in two bases, one for the known and one for the unknown cases [21]. The known cases are the ones used by the system to classify those tokens that are not new, i.e. tokens that have appeared in the training documents. The unknown base will be used for tokens that were not present in the training documents.

The main difference between the known and unknown cases is that in the former, the system saves the token itself, while in the latter a shape of the token is kept in order to allow the system to be able to classify unknown tokens by looking for cases with a similar shape. The shape of the token is given by a set of symbols: "A" and "a" for upper and lower case letters, respectively; "1" for numbers; "p" for stopwords; "g" for Greek letters; and "$" for highliting prefixes and suffixes. For example, "Dorsal" is represented by "Aa", "Bmp4" by "Aa1", "the" by "p", "cGKI(alpha)" by "aAAA(g)", "patterning" by "pat$a" ('$' separates the 3-letters prefix) and "activity" by "a$vity" ('$' separates the 4-letters suffix).

In the construction of cases, the training documents are read twice, once in the forward direction (from left to right), and once in the backward direction (from right to left). This is done in order to allow a more variety of cases and due to the fact that the decision of classifying a token as a gene/protein mention may be influenced by its preceding and/or following tokens. The known and unknown cases saved during the forward reading are used only for forward classification. In the same way, the backward cases are useful only to the backward classification.

In the testing step, the system searches the bases for the case most similar to the problem and the decision is given by the class of this case-solution. If more than one case if found, the one with higher frequency is the one chosen. The search procedure is separated in two parts, for the known and for the unknown cases and higher priority is always given to the known ones. The system considers both the forward and backward directions. This is important because sometimes some tokens are more easily recognized as part of a mention in only one of the two directions, especially for mentions composed of more than one token. For example, for the mention "cka1 delta cka2-8", the tokens "cka1" and "cka2" are easily recognized in any direction, forward or backward. However, "delta" would not be recognized as a mention in the forward direction, because in the training documents, "delta" usually appears as a non-mention token. However, in the backward direction, it may be recognized as a mention when preceded by another gene mention, as in the case of "cka2".

If no best known or unknown case is found in the search procedure, the token is classified as a gene/protein mention by default. This decision is due to the fact that if the token and its shape are both strange to the system, the possibility that it may be a mention is high, as its nomenclature usually includes sequences of letters, numbers and punctuations (such as hyphens, slashes, parenthesis, etc.). Our experiments have shown than this hypothesis is true in about 61% of the decisions. Table 3 (supplementary material) shows the mentions that were correctly and incorrectly classified as positive due to this assumption.

3.2 ML-Normalization

The normalization task is accomplished by ML-Normalization and consists of a flexible matching and a machine learning matching approaches that are currently available for four organisms: yeast, mouse, fly and human. Even if the system is initially implemented for only four organisms, it may be trained to support others for both matching approaches. To include new organisms, it is enough to provide their genome data available at Entrez Gene FTP[4] ("gene_info.gz" and "gene2go.gz" files).

ML-Normalization receives as input the original text and the mentions that have been extracted from it. However, it does not require the CBR-Tagger to be used as mention tagger. The initial lists of synonyms for the four organisms are the ones made available in the BioCreative challenges: BioCreative task 1B [2] for yeast, mouse and fly; and BioCreative 2 GN task [3] for the human.

The flexible matching is accomplished by a correspondence between the mentions extracted from the text and the synonyms of the organisms' dictionaries. It is considered flexible because both mentions and synonyms are previously pre-processed by means of applying some editing procedures for both, the mentions extracted by the taggers and the synonyms in the dictionaries. The editing operations are the same for all organisms. First the tokens are converted to lower case and separated according to symbols, punctuations, Greek letters and numbers. These subparts are then sorted alphabetically in order to avoid mismatching due to different ordering of the same tokens, as proposed in [15].

The system then performs a cleaning of the mention (or synonym) in order to remove parts that coincide with terms in BioThesaurus [22], stopwords or organism (NCBI Entrez Taxonomy). This is especially helpful for mentions composed of many tokens. The cleaning of the BioThesaurus terms is accomplished gradually, according to the frequency of the term in the lexicon, i.e., the terms with frequency higher than 10, 1000 and 10000, etc. By carrying out this gradual cleaning, we increase the possibility of finding an exact matching with no need to provide organism's specific information. Figure 1 illustrates an example of the editing procedure. All variations showed in bold are added to the dictionary (in case of a synonym) or considered in the matching procedure (in case of a mention). Table 6 (supplementary material) shows the effect of the gradual cleaning to the results of the normalization task.

In summary, one-token mentions or synonyms are filtered out if they coincide with BioThesaurus terms with frequency higher than 10, while the cleaning procedure is accomplished gradually as described above. The frequency threshold was experimentally determined (cf. supplementary material). The system also removes those mentions/synonyms composed of one character only, composed by no letters at all (only numbers and/or other symbols), or those that coincides with Roman numeral, Greek letters, amino acids, stopwords and organisms in NCBI Entrez Taxonomy. The final dictionaries of synonyms for the four refereed organisms are available for download at Moara download page.

The machine learning matching approach uses Weka [5] implementation of Support Vector Machines, Random Forests and Logistic Regression algorithms. In order to construct a training set for the algorithms, we used the methodology proposed by [23] in which the attributes of the training examples based on the comparison of two synonyms of the dictionary. When the comparison is between a pair of synonyms of the same gene/protein, it consists of a positive example for the machine learning

---

[4] ftp://ftp.ncbi.nih.gov/gene/DATA/

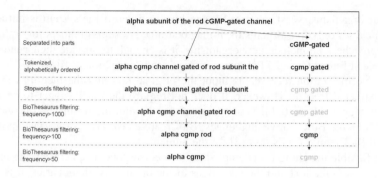

Fig. 1. Example of the editing procedure carried out for mentions and synonyms of the dictionaries. Variations are generated from the original text by isolating the tokens, ordering them alphabetically and filtering stopwords, and BioThesaurus terms gradually.

algorithm; otherwise, it is a negative example. One machine learning model is constructed for each of the organisms under consideration here. The features that represent the comparison of a pair of synonym are the following: indicative of equal prefix, equal suffix, equal number, equal Greek letter, bigram/trigram similarity, string similarity and shape similarity.

We have used Weka feature selection functionalities for analyzing the contribution of each of theses features. We have tried the feature selection methods based on chi-squared statistics and based on the gain ratio to measure the contribution of each attribute individually. The more meaningful features are the ones with higher scores (or gain) as presented in Table 5 (supplementary material).

During the testing step, the examples are built using the same features described above. We carry out the comparison of pairs of mentions and some of the synonyms of the dictionary, instead of the pairs of synonyms used in the training step. The output provided by the classifier is the indicative if the pair of mention-synonym is a match (positive class) or not (negative class).

In case where more than one identifier matches a given mention, a disambiguation strategy selects the best candidate by calculating the similarity between the provided text (e.g., the abstract) and a set of tokens representative of each of the candidates (gene-document). Each gene-document is automatically generated during the training step for each gene/protein of the organism by a compilation of some information extracted from several databases, such as SGD [24] for the yeast, MGI [25] for the mouse, FlyBase [26] for the fly and Entrez Gene [27] for the human. The fields collected for the construction of the gene-documents were symbols, aliases, descriptions, summaries, products, phenotypes, relationships, interactions, Gene Ontology terms related to the gene and their names, definition and synonyms. To include a new organism, the gene-documents are built using data from Entrez Gene [27] such as name, symbol, summary and the description of the GeneOntology [28] terms associated to the respective gene/protein.

The three scoring methodologies used for the disambiguation step are based on the original input text and each of the candidates' gene-document. The first of them uses the cosine similarity [29] while the second one counts the number of common tokens.

The gene-document with the highest cosine similarity or with the highest number of common tokens is chosen as the best candidate for the given mention, respectively. The third methodology is based on the other two and the decision is given by the higher product of the cosine similarity and the number of common tokens. Also, single or multiple disambiguation strategies are available. The first one selects only the best candidate (higher score) while the second one returns the top scoring ones according to a threshold that is automatically calculated as 50% of the value of the highest score candidate.

4   Results

Many experiments have been carried out during development in order to decide the final configuration of the system. These experiments have taken into account the taggers used to extract the mentions, the matching procedure (flexible or machine learning), features used in the machine learning algorithms and the disambiguation step. Additional results are available at the supplementary material page.

Table 1 presents the results for the gene/protein recognition problem, evaluated on the test dataset (5,000 documents) of the BioCreative 2 GM task [1], for the five training models of the CBR-Tagger, as well as the best result of the challenge. The mentions that have been extracted with all the five models are available for download at Moara download page. Table 4 (supplementary material) shows some comparative results for CBR-Tagger when using only backward or forward direction or when considering only the known or unknown cases.

Table 1. Results for the recognition task (test set) when comparing the training models

Tagger	Recall	Precision	F-Measure
CbrBC2	64.11	76.01	69.56
CbrBC2y	42.90	80.98	56.08
CbrBC2m	29.14	76.08	42.14
CbrBC2f	51.05	73.66	60.30
CbrBC2ymf	24.53	77.00	37.21
Best BioCreative	86.0	88.5	87.2

The results shown in Table 1 show that the CbrBC2 is the best dataset for training CBR-Tagger to the gene/protein recognition task. Table 7 (supplementary material) show results for the normalization task, evaluated on the development corpora of BioCreative task 1B [2], for the yeast, mouse and fly, and on the BioCreative 2 [3] for the human. The mentions have been extracted with a mix of taggers that include the five models of CBR-Tagger as well as ABNER [11] and Banner [12]. These experiments show that a tagger trained with specific organism's documents may improve the precision and f-measure and the combination of more than one tagger may improve the recall.

Table 2 presents the results for the gene/protein normalization task, evaluated on the test corpora of BioCreative task 1B [2], for the yeast, mouse and fly, and on the BioCreative 2 [3] for the human, as well as the challenge best results. Mentions were extracted with the mix of CbrCB2ymf, ABNER and Banner. Many experiments have been carried out on the development dataset in order to achieve the best set of parameters that works reasonably well for the four organisms. Experiments include the evaluation of machine learning algorithm (Support Vector Machines) and their set of features, disambiguation procedure using single selection and the score based on the product of cosine similarity and the number of common tokens. Additional results are presented in the supplementary material page, such as a comparison of the string similarity feature of the machine learning algorithms (Table 8), the different disambiguation strategies (Table 9) as well as details on the errors (true positives, false positives, false negatives) for the flexible matching (Tables 10) and machine learning matching (Table 11).

Table 2. Results for the normalization task (test set) when comparing the flexible and the best machine learning matching approaches

Organism	Best BioCreative			Exact matching			ML matching		
	R	P	FM	R	P	FM	R	P	FM
Yeast	89.4	95.0	92.1	83.52	95.17	88.97	84.34	81.67	82.99
Mouse	81.9	76.5	79.1	77.57	65.83	71.22	79.60	32.90	46.56
Fly	80.0	83.1	81.5	69.76	59.12	63.58	69.00	55.22	61.35
Human	83.3	78.9	81.0	83.31	55.00	66.26	85.99	29.13	43.52

5 Discussion

In this work we present a novel methodology for the gene/protein mention problem based on a classification approach using case-based reasoning. Results show the suitability of this approach for the gene/protein recognition problem, which is a necessary step in many text mining procedures in biomedicine. Although the results presented for the gene mention extraction seem to indicate that training systems with specific documents might result in a worse performance, the results presented for the normalization task clearly shows that it is not the case. We consider the gene/protein recognition as a preceding step for the normalization problem, and the improvement of the latter is the main goal of a gene/protein tagger.

The methodology proposed here is also feasible for the complex gene normalization task due to the fact that a satisfactory F-Measure is obtained with no need of making adaptations in the algorithms to fit any particular organism. In addition to this, our system has been designed with very little dependency with custom dictionaries or annotated documents, which are generally not publicly available, and no specific knowledge was inferred from experts, as some other systems do [13].

When comparing our results with those reported in the two editions of BioCreative we have found that those that have achieved better F-Measure than ours have made use of an organism-specific procedure either for a curated dictionary or for the matching strategy. Even if these approaches produce good results for a specific organism, they cannot be extended to new organisms without a similar set of rules inferred from expert knowledge. Therefore, using or reproducing those existing methods with new organisms is very time consuming and sometimes impossible. In our system, we use only publicly and general available information for every organism. It is true that we can not exclude some of the needed organism-specific information like the dictionary of synonyms or gene/protein annotations which are necessary for the matching and disambiguation tasks, respectively. However, this information can be obtained from public databases and no organism-specific tailoring is necessary to obtain satisfactory results.

By analyzing the errors from the development dataset, some of the false negative mistakes are still due to mentions that could not been extracted from the taggers. Also, a high number of false negatives are due to a wrong disambiguation, with the consequent generation of many other false positives. Results provided with this approach look very promising and can certainly have more room for improvement, in particular with the mention extraction and disambiguation procedures. These two procedures were not originally in the main focus of this study although results clearly indicate that more efforts should be devoted to those since global improvement heavily depends on their performance.

The final configuration of the system may be tailored by the user according to its needs, in order to achieve a best precision, recall or F-Measure. Moara is freely available to be used by the scientific community and includes classes that allow the user to test the tagger and the machine learning normalization described here, including the possibility of choosing the training documents for the tagger and the features used in the machine learning matching as well as the disambiguation strategy. A UIMA wrapper for the functionalities described here is under development in order to integrate Moara to U-Compare [30].

Acknowledgements. This work has been partially funded by the Spanish grants BIO2007-67150-C03-02, S-Gen-0166/2006 and PS-010000-2008-1. The authors acknowledge support from Integromics, S.L.

References

1. Smith, L., et al.: Overview of BioCreative II gene mention recognition. Genome Biology 9 (Suppl. 2), S2 (2008)
2. Hirschman, L., et al.: Overview of BioCreAtIvE task 1B: normalized gene lists. BMC Bioinformatics 6(Suppl.1), S11 (2005)
3. Morgan, A.A., et al.: Overview of BioCreative II gene normalization. Genome Biology 9(Suppl. 2), S3 (2008)
4. Aamodt, A., Plaza, E.: Case-Based Reasoning: Foundational Issues, Methodological Variations, and System Approaches. AI Communications 7(1), 39–59 (1994)
5. Witten, I.H., Frank, E.: Data mining: Practical machine learning tools and techniques, 2nd edn. Morgan Kaufmann, San Francisco (2005)
6. Cohen, W.C., Ravikumar, P., Fienberg, S.E.: A Comparison of String Distance Metrics for Name-Matching Tasks. In: II Web Workshop on International Joint Conference on Artificial Intelligence, Acapulco, Mexico (2003)
7. Fukuda, K., et al.: Toward Information Extraction: Identifying protein names from biological papers. In: Pacific Symposium on Biocomputing (PSB 1998), Hawaii, USA (1998)
8. Finkel, J., et al.: Exploring the boundaries: gene and protein identification in biomedical text. BMC Bioinformatics 6(Suppl. 1), S5 (2005)

9. McDonald, R., Pereira, F.: Identifying gene and protein mentions in text using conditional random fields. BMC Bioinformatics 6(Suppl. 1), S6 (2005)

10. Zhou, G., et al.: Recognition of protein/gene names from text using an ensemble of classifiers. BMC Bioinformatics 6(Suppl.1), S7 (2005)

11. Settles, B.: ABNER: an open source tool for automatically tagging genes, proteins and other entity names in text. Bioinformatics 21(14), 3191–3192 (2005)

12. Leaman, R., Gonzalez, G.: BANNER: an executable survey of advances in biomedical named entity recognition. In: Pac. Symp. Biocomput., pp. 652–663 (2008)

13. Fundel, K., et al.: A simple approach for protein name identification: prospects and limits. BMC Bioinformatics 6(Suppl.1), S15 (2005)

14. Crim, J., McDonald, R., Pereira, F.: Automatically annotating documents with normalized gene lists. BMC Bioinformatics 6(Suppl.1), S13 (2005)

15. Liu, H., Wu, C., Friedman, C.: BioTagger: A Biological Entity Tagging System. In: BioCreAtIvE Workshop Handouts, Granada, Spain (2004)

16. Hakenberg, J., et al.: Inter-species normalization of gene mentions with GNAT. Bioinformatics 24(16), 126–132 (2008)

17. Xu, H., et al.: Gene symbol disambiguation using knowledge-based profiles. Bioinformatics 23(8), 1015–1022 (2007)

18. Farkas, R.: The strength of co-authorship in gene name disambiguation. BMC Bioinformatics 9, 69 (2008)

19. Neves, M., et al.: CBR-Tagger: a case-based reasoning approach to the gene/protein mention problem. In: BioNLP 2008 Workshop at ACL 2008, Columbus, OH, USA (2008)

20. Neves, M.: Identifying Gene Mentions by Case-Based Reasoning. In: Second BioCreative Challenge Evaluation Workshop, Madrid, Spain (2007)

21. Daelemans, W., et al.: MBT: A Memory-Based Part of Speech Tagger-Generator. In: Fourth Workshop on Very Large Corpora., Copenhagen, Denmark (1996)

22. Liu, H., et al.: BioThesaurus: a web-based thesaurus of protein and gene names. Bioinformatics 22(1), 103–105 (2006)

23. Tsuruoka, Y., et al.: Learning string similarity measures for gene/protein name dictionary look-up using logistic regression. Bioinformatics 23(20), 2768–2774 (2007)

24. Cherry, J.M., et al.: SGD: Saccharomyces Genome Database. Nucleic Acids Res. 26(1), 73–79 (1998)

25. Eppig, J.T., et al.: The Mouse Genome Database (MGD): from genes to mice–a community resource for mouse biology. Nucleic Acids Res. 33(Database issue), D471–D475 (2005)

26. Gelbart, W.M., et al.: FlyBase: a Drosophila database. The FlyBase consortium. Nucleic Acids Res. 25(1), 63–66 (1997)

27. Maglott, D., et al.: Entrez Gene: gene-centered information at NCBI. Nucleic Acids Res. 35(Database issue), D26–D31 (2007)

28. Ashburner, M., et al.: Gene ontology: tool for the unification of biology. The Gene Ontology Consortium 25(1), 25–29 (2000)

29. Shatkay, H., Feldman, R.: Mining the biomedical literature in the genomic era: an overview. J. Comput. Biol. 10(6), 821–855 (2003)

30. Kano, Y., et al.: U-Compare: share and compare text mining tools with UIMA. Bioinformatics (2009)

Author Index

Printing: Mercedes-Druck, Berlin
Binding: Stein+Lehmann, Berlin